WINGS IN THE SEA

WINGS IN THE SEA
THE HUMPBACK WHALE

LOIS KING WINN & HOWARD E. WINN

PUBLISHED FOR UNIVERSITY OF RHODE ISLAND
BY UNIVERSITY PRESS OF NEW ENGLAND
HANOVER AND LONDON, 1985

UNIVERSITY PRESS OF NEW ENGLAND

BRANDEIS UNIVERSITY	UNIVERSITY OF CONNECTICUT	UNIVERSITY OF RHODE ISLAND
BROWN UNIVERSITY	DARTMOUTH COLLEGE	TUFTS UNIVERSITY
CLARK UNIVERSITY	UNIVERSITY OF NEW HAMPSHIRE	UNIVERSITY OF VERMONT

Printed in the United States of America

LIBRARY OF CONGRESS CATALOGING IN PUBLICATION DATA

Winn, Lois King, 1944–
  Wings in the sea.

  Bibliography: p.
  1. Humpback whale.   I. Winn, Howard Elliott, 1926–
II. Title.
QL737.C424W55   1985   599.5'1   84–40598
ISBN 0–87451–335–9
ISBN 0–87451–336–7 (pbk.)

For our children Brett, Bill, Matthew,
Carole, Greg, and Eric,
and to the memory of Brandon

CONTENTS

ACKNOWLEDGMENTS

Without the contributions of the many people mentioned throughout the book, we would not have been able to tell the humpback's story. We would especially like to thank Paul Perkins, former graduate students Richard Edel, James Hain, Gerald Scott, Bill Steiner, Algis Taruski, Tom Thompson, and Marilyn Nigrelli, and Ida King for their many contributions and help over the years.

For taking the time to share information with us we are grateful to Horace Beck, Tim Burgess, Carole Carlson, William Dawbin, Raymond Duguy, Capt. Johnny Harms, Herbert Hays, Louis Herman, James Hudnall, Åge Jonsgård, Charles Jurasz, Finn Kapel, Steve Katona, Greg Kaufman, Robert Kenney, Steve Leatherwood, Stormy Mayo, James Mead, Ken Morse, Carl Mrozek, George Nichols, Masaharu Nishiwaki, Catherine Osborne, Charlie Potter, Bill Schevill, Heimir Thorleifsson, Michael Tillman, Frank Watlington, and Hal Whitehead, plus others whose names were too numerous to list. Special thanks go to William C. King for illustrating the book.

Our gratitude also goes to the captains, crews, and scientific parties of the R/V *Trident*, *Sir Horace Lamb*, *Eaglet II*, *Stormy Petrel*, *Osprey*, *Sea World*, *Gypsies Five*, *Chewaucan*, *Kapduva*, *Sway*, and the R/V *Tioga*.

Support for research over the years was supplied by the Office of Naval Research, the Center for Field Research (Earthwatch), the Oceanic Society, the American Philosophical Society, the Marine Mammal Commission, and the Minerals Management Service. Many of the photographs were taken during research contracts with the Office of Naval Research and the Bureau of Land Management (now the Minerals Management Service). We also acknowledge the help of the government officials of Tonga, the Northern Marianas, Mexico, the Dominican Republic, and the Cape Verde Islands.

Permission to use photographs was given by John Arsenault; Jane Beck; Ann Brayton; Gary Carter; William Dawbin; Jeff Goodyear; James Hain; Charles Jurasz; Robert Kenney; Geoffrey LeBaron; National Marine Fisheries Service, Marine Mammal Division, Seattle; Carol Price; Smithsonian Institution; Cetacean Research Program, Provincetown, Massachusetts (Charles Mayo); Michael Williamson; and David Woodward. Since most University of Rhode Island photographs were taken under grants, the photographer is not recorded.

ACKNOWLEDGMENTS

Raoul Millais gave us permission to use two plates from *The Mammals of Great Britain and Ireland* by J. G. Millais, and permission to quote from *Follow the Whale* by I. T. Sanderson was obtained from Sabina W. Sanderson. Passages from *The Quest of the Schooner Argus* by Alan Villiers are reprinted with the permission of Charles Scribner's Sons, copyright 1951, 1979 by Alan Villiers.

L.K.W.
H.E.W.

WINGS IN THE SEA

INTRODUCTION

The humpback whale has existed for about ten million years. Some whale species have become extinct, and others no longer exist in certain parts of the world. The humpback has survived climatic changes and a relentless pursuit by humans, and although its numbers are dangerously reduced in some areas, it can still be found in all the oceans of the world. The humpback is the only whale that during some part of the year can be found off almost every coastline or island around the world. Why has the humpback survived? It may be that it is more adaptable than other whales.

The humpback is unique among whales. Although it is large, averaging about 50 feet in length, it cannot compare with the largest animal to ever live on the earth, the 95-foot blue whale. Yet its physical appearance is noteworthy. I. T. Sanderson, in his book *Follow the Whale* (1956), wrote that the humpback, "although basically cetacean, defies description." No other whale has various knobs and bumps on its snout, nor do any others have pectoral fins up to fifteen feet long. These unusual fins enable the humpback to engage in maneuvers and acrobatics impossible for other whales, and to move beneath the water with the grace and agility of a bird in flight.

While on its tropical mating and calving grounds the humpback also produces a hauntingly beautiful song unlike the sounds of other whales. Discovered in 1969, their song has become the symbolic voice of all whales. The song has accompanied symphonies and recording artists all over the world, and as a representative natural earth sound it is traveling through space aboard Voyagers I and II.

We were first introduced to the humpback whale in 1969 while on a research cruise in the West Indies. Since that first cruise we and other researchers at the University of Rhode Island have lived with and studied humpbacks throughout the North Atlantic and in the North and South Pacific. Portions of the book deal with some of the experiences we have had with the humpbacks and the results of our research. When we began there were very few scientists studying whales. As people became aware of the whale's endangered status, and photographers and scientists began to publicize the lives of these mysterious animals, more funds became available for research. In a few short years humpbacks were being studied on their feeding and breeding grounds in many areas throughout the world. The information obtained in

these studies has added significantly to our understanding of the humpback whale and its life in the sea.

One of the problems with studying whales at sea is that much of their time is spent beneath the water, offering researchers only brief glimpses of their activities. Each new bit of information must be fitted with others like the pieces of a puzzle, and the resulting information gaps must be filled in by speculation. With whales especially, a great deal of speculation has often been presented as fact. We caution the reader that any theories or guesses presented in this book are just that and will have to be verified by future studies.

When we decided to write a book about the humpback, we wanted it to be as complete and accurate as possible. We turned to the scientific literature and discovered that some early scientists and naturalists made important and accurate contributions that unfortunately are overlooked today. In the extensive bibliography that we offer in lieu of footnotes we have tried to include the work of most of the researchers who have made significant contributions to the study of the humpback's natural history. Since we wanted to present the story of the humpback from every aspect of its recorded history, we also examined folklore, myths, and the logbooks and journals of whalers, explorers, seamen, and naturalists from around the world for clues to the humpback's past. At first the humpback was regarded with awe, fear, and superstition; later, as men became more knowledgeable and skilled, the humpback became a source of food and oil. The humpback led whalers from the North Atlantic to all the oceans of the world and the whale to the brink of extinction.

Although we know a great deal about the humpback, many questions remain. In the following chapters we tell the humpback's story as it stands today, leaving the unanswered questions for tomorrow. Perhaps the final chapter of the humpback's story will never be written.

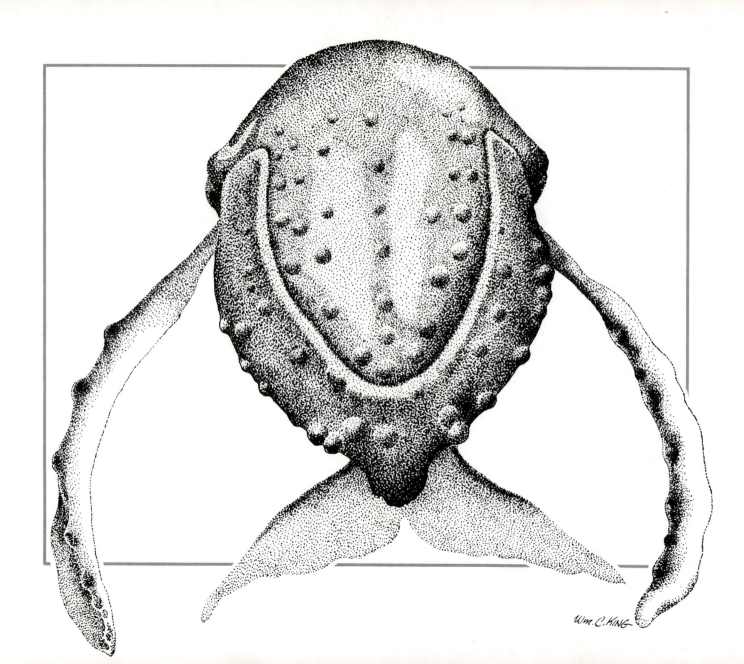

1. MEGAPTERA

Thirty-four million years ago, during the Oligocene epoch, the North Atlantic was slowly formed as Europe separated from North America. The climate was warm, and much of the North Atlantic was subtropical and tropical. The toothed whales had already been in existence for twenty million years. They evolved rapidly, feeding on the abundant supply of fish in the oceans.

The first baleen or mysticete whales emerged late in this period. Although the mysticetes are descendants of ancestors with teeth, they are not directly related to the modern toothed whales, the odontocetes. It is believed by some that the toothed and baleen whales each developed independently from different land ancestors. The early forms of toothed mysticetes at some point began to develop hardened ridges of skin in their mouths, much the same as those found inside the mouth of a dog. Eventually the ridges evolved into baleen plates, from which grew the hairy fringes that enabled these whales to strain smaller animals from the sea. Now the baleen whales could take advantage of a vast new food supply. From fossil remains we know that a primitive group of mysticetes, the Cetotheriidae, were much smaller than the baleen whales living today, averaging only between 9 and 33 feet long.

During the next epoch, the Miocene, the earth began to cool. By the beginning of this epoch, 25 million years ago, the continents had separated further. The Atlantic had become larger, India had joined Asia, and most important, Australia had separated from Antarctica, setting up the Antarctic circulation. Areas of upwelling were created as they are today, where two water masses move apart and the diverging surface waters are replaced by deeper water rich in nutrients. Phytoplankton develop, then the zooplankton that feed on them, and finally the fish and squid that feed on the zooplankton; a food chain is established. Mysticetes evolved rapidly in these areas of upwelling, where they found food in abundance. This period gave rise to the most primitive of the baleen whales that are still living today, the right whale and the bowhead whale, members of the family Balaenidae.

No one can be positive where the first humpback whales developed, but it may have been in the Atlantic west of southern Europe and northern Africa, where upwelling occurs. At any rate, about ten million years ago, in a tropical shallow area, the first whale evolved with humpback characteristics.

The humpback whale is a member of the family Balaenopteridae, more commonly called rorquals. Rorquals are relatively fast swimmers with pleated folds of skin (ventral grooves) running from their chin to their midsection. In the late nineteenth century fossils from the Pliocene epoch (seven million years ago) were discovered near Antwerp, Belgium, and in eastern Holland. At this time rorquals were much smaller than those living today, averaging 10 to 50 feet. Only during the last few million years have rorquals attained their current size of 30 to 100 feet. Included in the rorqual family are the blue whale, the fin whale, the sei whale, Bryde's whale, and the Minke whale in the genus *Balaenoptera*. The humpback is the lone representative of its genus, *Megaptera*. Unlike other rorquals, it developed a complicated call or song, perhaps because of competition from related humpback types now extinct or for some other functional reason such as sexual selection (wherein males with the most complicated or novel songs are more successful at mating).

As the seas continued to cool, the humpback had to leave the tropics and travel to colder waters, where food was more abundant. Whales developed blubber to keep them warm in the cold of the Arctic and Antarctic, where they fed during the summer months, and to use as a food supply while in the tropics. But each winter they had to return to the tropical seas of their origin to mate and give birth, beginning a migration that continues to this day.

Today humpbacks are found in all the oceans of the world. In the polar areas they divide themselves into several stocks. Although there is some evidence of interchange between stocks while they are in the colder seas, this does not seem to be a common occurrence. When migration to the southern breeding grounds begins, the stocks divide themselves into smaller groups, each of which follows traditional routes to its mating and calving ground. One of the longest routes is taken by an Antarctic stock of humpbacks, which migrates to the Gulf of Panama on the west coast of South America, a distance of 4,000 miles. There may be some intermingling between the substocks while they are on the mating grounds, particularly in some areas; but most individuals probably return to the same area in the tropics year after year. In tropical areas where upwelling occurs, such as off the coasts of Dakar, Africa, and Costa Rica and in areas in the Arabian Sea, food is available to the humpbacks all year. In these locations it is not necessary for humpbacks to migrate, and some appear to stay throughout the year.

No one knows when man first encountered the humpback whale, but it was known to sailors and naturalists long before it was identified and named by scientists. Past glimpses of the humpbacks have come down to us in various writings. One of the first books to mention whales was an account of a voyage made by the Greek sailor Nearchus. In 326 B.C. Nearchus was instructed by Alexander of Macedon, who was then on campaign in present-day India, to construct a fleet and sail back to

Mesopotamia along the Makran coast (in present-day Pakistan), an area where humpbacks and some other large whales occur. According to the account of the voyage, Nearchus and his men encountered a great number of whales, which caused a commotion in the sea and "raised so dense a mist by their blowing that the sailors could not see where they stood." The pilots informed the sailors that these animals would leave quickly if they blew trumpets and clapped their hands. Nearchus headed straight for the whales, frightening them with the trumpets. The whales dived, then rose again at the bows of the vessels, "so as to furnish the appearance of a sea fight, but they soon made off." Although we cannot be sure Nearchus saw humpbacks, they are still abundant in this area of the Arabian Sea.

As the humpback became more familiar to seamen, it was given various names referring to its humped back (see Fig. 1) or the round bumps on its head. To the French it was *baleine a bosse* (whale with hump), to the Germans *Pflockfisch* (peg fish), and to the Norwegians *knolhval* (knobbed whale). The unusual antics of the humpback inspired the Russians to choose the name *vessyl kit*, or merry whale. In the western provinces of Japan, blind men carrying lutes on their backs and chanting the Jishinkyo or "sutra of the earth goddess," once went from house to house to honor the deity of the hearth. To the whalemen of Japan, the bunch, or protuberance, formed by the humpback's low dorsal fin resembled one of these blind men shouldering his lute case, hence the name *zatokujira*, or "whale of the blind people."

In the eighteenth century Carolus Linnaeus, a Swedish naturalist and botanist, devised a method for naming plants and animals. In his classification system each living thing was given two names in Latin, the first for the group (genus) and the second for the species; some names of both types refer to distinctive characteristics of the animals or plants. For a name to be adopted by the scientific community it must be accompanied by an accurate description of the animal. Many of the early names applied to the humpback were not acceptable because the descriptions were confusing, sometimes containing characteristics of two or more different species.

The name *Balaena boops* (ox-eyed whale) was applied to the humpback by O. Fabricius in 1780, but since Linnaeus had already assigned this name to a fin whale, it could not be used for the humpback. In 1781, G. H. Borowski, a German zoologist, described a whale found in abundance off the coasts of New England and named it *Balaena novaeangliae* (New England whale). This name for the humpback was overlooked, however, and on the basis of a description given by the British whale naturalist P. Dudley, A. Bonnaterre, a French zoologist, in 1789 proposed the name *Balaena nodosa* (knobby whale).

Until 1846 all whalebone whales were considered part of the genus *Balaena*. In that year J. Gray, a British zoologist, pronounced the humpbacks a separate group and

Fig. 1. The humpback gets its name from the characteristic dive shown here. The whale lowers its head and arches its back, and as the dive is completed the flukes come out of the water. This occurs only when the humpback is making a deep dive. On shallower dives the flukes do not come out of the water.

changed their genus designation *Balaena* to *Megaptera*. The name *Megaptera nodosa* was thereafter used to refer to the humpback until 1932, when Remington Kellogg, an American whale biologist, showed that the description given by G. Borowski was valid. Since then *Megaptera novaeangliae* has been the formal designation of the humpback whale. *Megaptera* means "large-winged" and refers to the humpbacks' long pectoral fins; as we have seen, *novaeangliae* refers to New England, where Borowski first described the humpbacks that were found in great numbers there.

The humpbacks on the European side of the Atlantic were not scientifically recognized until 1829, when Rudolphi described a specimen that had stranded at the mouth of the Elbe River in Germany in 1824 and proposed the name *Balaena longimana* (long-armed whale). As new herds of humpbacks around the world were discovered they were given new scientific names, since it was assumed that they were different species. Eventually there were 23 names in the scientific literature referring to the humpback whale. Today all humpbacks are assumed to be a single species, *Megaptera novaeangliae*.

During this period of scientific revival there were many unqualified people writing books about whales. Most of them got their information second hand; some had never even seen a whale. Thus Henry Dewhurst's *The Natural History of the Order Cetacea*, published in 1834, describes whales seen during a voyage to Greenland. One of Dewhurst's more imaginative creatures, the "Physeter Gibbosa," bears some resemblance to the humpback but is obviously a composite of several species:

The *Physeter Gibbosa* is a native of the Northern Seas, and is said to have the same general form as the *Balaena mysticete*, with the exception of its being of much smaller dimensions, and having the back furnished with one or more tubercles, which have been denominated bunches, and hence gave origin to its common appellation. A variety of this species is found on the coasts of New England; and another, having six tubercles along the back was supposed by the late Dr. Shaw to inhabit the coasts of Greenland, but neither of these varieties appears to be very accurately known either by the whalers or zoologists. Their baleen is said to assume a pale or whitish color, and in the lower jaw they resemble the Cachalots, by having teeth.

Whoever told Dewhurst that a whale existed with baleen in its upper jaw and teeth in the lower either was totally ignorant about whales or was having a bit of fun with him. In the book he even included a drawing of this strange creature, obviously a combination of the humpback, the sperm whale, and perhaps the scragg whale. (Scragg whales were often mentioned by the early whalers of New England, but no one is quite sure what they were. Some scientists believe they may have been the last of the Atlantic gray whales, which are now extinct.)

Today we know a lot more about the physical characteristics of humpback whales (see Fig. 2). Like humans they come in various sizes, shapes, and colors. One of

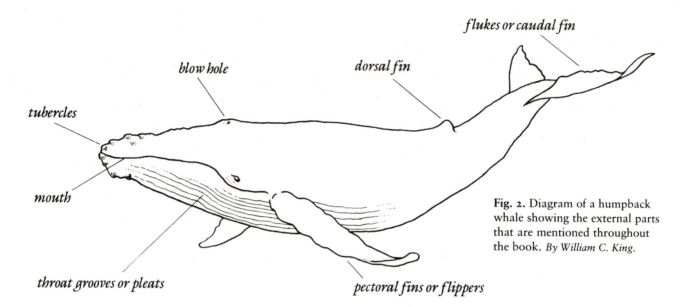

flukes or caudal fin

dorsal fin

blow hole

tubercles

mouth

throat grooves or pleats

pectoral fins or flippers

Fig. 2. Diagram of a humpback whale showing the external parts that are mentioned throughout the book. *By William C. King.*

the most distinctive features of the humpback is its pectoral fins. The fin length, although highly variable, is usually about one-third the total body length. No other whale has pectoral fins this long. The wavy front edge of the flipper carries nine to ten tubercles or bumps protruding at the joints and phalanges, and the bones in the flipper are identical to those in a human arm and hand. Flipper shapes vary from animal to animal; some are long and narrow, others short and wide, and others intermediate. Short curved folds are present at the base of the flipper, some extending onto the flipper and some running above the base. One to three furrows are found at the corner of the mouth.

The average length of a northern humpback is 45 to 52 feet; humpbacks in the southern hemisphere are bigger, averaging 59 feet. It may be southern hemisphere whales spend less time feeding, in which case a larger body size would provide more blubber to live off while the whale is in tropical waters. Another possibility is that food resources are greater in the Antarctic than in the Arctic. Female humpbacks average a foot or two longer than males, perhaps because they need the extra size to provide nourishment to their fetuses and young calves. The largest humpback ever recorded was an 88-foot female taken at Bermuda. Although there is some doubt about the authenticity of this measurement, the lengths of other body parts were in proportion to an animal of that size. Nearchus reported the size of the humpbacks in the Persian Gulf as 74 feet, a realistic estimate for that time. If the larger whales were killed, the genes for a smaller size may have become dominant. It could also be that an increase in competition for food has limited the potential size that humpbacks can attain.

A humpback's head is large in proportion to the rest of its body. In rows along the head and lower jaw are found numerous round knobs, each about as large as half an orange and usually containing a single gray hair from 3/4 inch to 1 1/4 inches long. Although some of the knobs lack hairs, they retain the hair sac. When the whale is viewed from above, the distribution of these knobs is not always symmetrical.

The snout is blunt, and the upper jaw is much narrower than the lower. Beneath the chin is an irregularly shaped knob that seems to enlarge with age. The eye is dark brown with a kidney-shaped pupil (see Fig. 3), and the ear is a flattened slit located just behind and below the eye. This opening is very difficult to find, since it is only about ¼ inch long.

Fig. 3. The eye of the humpback whale. This whale is looking back toward its tail. *Photo by Gary Carter.*

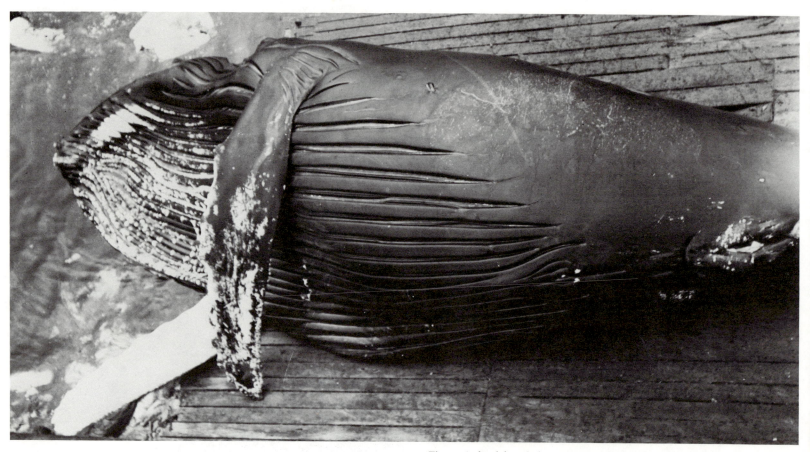

Fig. 4. A dead female humpback, showing the ventral grooves or pleats, which run from the chin to the navel. The pleats expand when the whale is feeding, allowing it to take in great quantities of water and food. *Photo courtesy of Smithsonian Institution.*

Folds of skin called ventral pleats or grooves start at the chin and run parallel to each other, reaching to the navel or just slightly beyond (see Fig. 4). Some of the pleats fuse together and in the region of the chin may converge to the mid-line. The pleats of the humpback are wider, deeper, and fewer than those of other rorquals. Considerable variation can be found in their number, and females seem to have more than males. Perhaps this allows females to get more food, since more pleats means a larger pouch. North Atlantic animals have from 14 to 22 pleats, those of the North Pacific from 12 to 26.

The flukes or tail fins are deeply notched in the center and usually serrated along the rear margin. Since these serrations are present in the embryo, they are not the result of injury as was once thought. The dorsal fin is short and thick, usually not exceeding twelve inches in height. This fin can be pointed, sickle-shaped, or blunt.

Three main color patterns occur in humpbacks throughout the world. One kind of humpback is completely or almost completely black, with a few small speckles; the flippers are dusky above, occasionally with an exceptionally white border. The second kind is black above with white underneath from the chin to the tail. The third kind has a piebald belly. On the underside of this third kind black dots, spots, and speckles create intricate and beautiful abstract patterns, contrasting sharply with the creamy white skin. Some whalers believed that a humpback's color changed as it aged, with younger animals having more white than older animals. This does not appear to be true, but there is still some question whether the various stocks of humpbacks have distinctive color patterns.

The back and flanks of humpbacks are black, occasionally with a brownish tinge but always darker than other rorquals. Flipper color is highly variable: the top can be black, spotted, marbled, piebald, or completely white. Lower surfaces are usually uniformly white, sometimes with spots or speckles near the base, occasionally gray and very rarely black. Ninety percent of the humpbacks in the North Atlantic have mostly white flippers; in the southern hemisphere black flippers are common. The flukes are black above, the lower side ranging from all white to all black with all patterns of black and white in between.

Contrary to our expectations, the logs and journals of early explorers did not yield many stories about the unusually long flippers of the humpback. One exception we found may explain why these unique appendages were not often mentioned. In *Ramusio's Voyages*, the Spanish explorer G. Oviedo wrote a chapter on the whales found off the coast of Spanish America.

I will relate what I myself with many others saw in the mouth of the Gulf of Orotingna, which is 200 leagues distant from the town of Panama toward the West. In 1529, going out of the Gulf into the open sea, to go to the town of Panama, we saw at the mouth of the Gulf a fish or marine animal extremely large, and which from time to time raised itself straight out of the water. And that which was to be seen above the water, which was only the head and two arms, was considerably higher than our caravel with all its masts. And being elevated in that way it let itself fall and struck the water violently, and then after a little time returned to repeat the act, but not, however, throwing up any water from the mouth, although in falling down with the blow and the fall it made much water rise up into the air. And a cub of this animal, or one like it but much smaller, did the same, deviating always somewhat from the larger one. And from what the sailors and others who were in the caravel said they judged it to be a whale, and the smaller a whale's cub. The arms which they showed were very large, and some have said that the whale has no arms. But the one which I saw, was of the manner I have said, for I went with the others in the caravel, where came also Father Lorenzo Martion, canon of the church of Castiglia dell'Oro; and the pilot was John Cabezas; and with us came also a gentleman named Sancio di Tudela, and many others, who are alive and can testify the same thing, because I would never wish to speak of such things without witnesses.

It seems that Oviedo was willing to attest to such a sight only because he had many witnesses, including a priest. His description continues:

By estimate, and as it seemed to me, each arm of this animal might be 25 feet long and as thick as a barrel and the head more than 14 or 15 feet long, and very much thicker and the rest of the body more than as much again.

It raised itself up and that which it showed in height was more than five times the height of a middle-sized man, which makes 25 feet. And the fear was not a little that all had when

Fig. 5. Humpbacks are home for many sea creatures. Here stalked barnacles grow from acorn barnacles attached to the tubercles of the humpback. The whale lice on the first tubercle also make their home on the whale. *Photo courtesy of Smithsonian Institution.*

with its leaps it came alongside our vessel, because our caravel was small. And from what we could surmise it seemed that this animal felt pleasure, and made holiday of the weather which was approaching; for soon there arose in the sea a strong west wind, which was much to our advantage.

This account may well be the first description of the humpback whales that mate and breed off the west coast of Panama.

Unlike most other whales, humpbacks are usually heavily infested with external parasites (see Fig. 5). The Greenland whalers believed they were born with the pests. I. T. Sanderson, in his book *Follow the Whale*, gives a very apt description of the beasts that make their home upon the humpback:

Huge sessile barnacles and masses of stalked barnacles festoon their sides and cluster in all cracks, crevices, folds and other interstices. Vast "lice," actually, strange, almost bodiless, but many-legged creatures known as pycnogonids (they are amphipods), meander all over them by hooking their sharp claws into the skin, and myriads of lesser, more numerous-legged crustaceans scurry hither and yon about their person. Certain molluscs bore into their blubber, just as they do driftwood, rock, or even concrete; plume-headed worms either sink their gelatinous bodies into the poor beast's flesh or construct convoluted calcareous tunnels on its surface; and only Nature knows what other freebooters bum rides on the patient beast's exterior or in its inner recesses.

The two most common pests found on the humpback are the stalked and acorn barnacles, the latter being the type most commonly found on the bottom of ships and on pilings along the shore. They attach themselves directly to the skin of the whale, often embedding themselves to a depth of four inches. The stalked barnacle attaches itself to the acorn barnacle. One scientist reported examining a humpback that was carrying over a thousand pounds of stalked and acorn barnacles. The lower jaw and the darker parts of the belly sometimes bear small white circles; some are probably scars from acorn barnacles that have become detached. Some could be wounds left by lampreys, eel-shaped fish that fasten themselves to other aquatic animals. Others may be scars from the bite of the "cookie cutter" shark (*Isistius brasiliensis*), a fish found in warmer waters. Most humpbacks are variously scratched and scarred, possibly as a result of contact with the barnacles of others.

It is quite possible that humpbacks itch. Many observers have seen humpbacks in shallow water apparently attempting to rub away some of the pests embedded in their hides. F. D. Ommanney, in his 1971 book *Lost Leviathan*, reported that they were frequently seen rubbing themselves against rocks. According to the English naturalist John Millais, in his 1906 book *The Mammals of Great Britain and Ireland*, Norwegian whaling captains told him the humpback was the only species of large whale that voluntarily came into shallow water and rubbed their heads and pectoral fins against rocks to rid themselves of barnacles. Two captains reported seeing humpbacks rubbing their heads against rocks so close inshore that a stone thrown from shore could have hit them.

Artists whose portraits of the humpback depict a stout, clumsy-looking animal can never have seen a living humpback, much less observed one underwater. Their models must have been dead animals, bloated with gases, stiff with death. Even today some old illustrations are copied that amount to a caricature of this graceful animal. Not until the first underwater photographs were taken was the sleek gracefulness of the humpback fully revealed. Underwater the bulk becomes streamlined; the huge muscles ripple just beneath the surface of the skin. Flippers, constantly in motion, bend and flex gracefully, giving the whale an amazing agility for its huge size (see Fig. 6).

Now we have a physical picture of the humpback. But what sort of an animal is it: friendly or fierce, timid or bold? According to Icelandic folklore, the humpback was a friend of man. When sailors' frail barks were sur-rounded by the ferocious and carnivorous whales of the north, the humpback would endeavor to rescue its friends from danger and accompany them back close to shore. Whalers and naturalists of the late 1800's and early 1900's had varying opinions of the humpback's character. In 1859 the author and seaman Charles Nordhoff called the humpback "the most stupid of whales." A Norwegian whaling captain named A. Ingebrigtsen disagreed: "The humpback is far more intelligent than other species of whale." To Millais it was "a gay sportive animal"; and Frank Bullen, another nineteenth-century author and seaman, observed, "There be few creatures in earth, air, or sea that lead a happier life, or enjoy it with a greater zest than the humpback."

Where are humpbacks found? How many of them are there? What are their patterns of feeding, migrating, and breeding? To these questions we turn next.

Fig. 6. The underwater grace and agility of the humpback whale are apparent in this photograph. *Photo by Dave Woodward.*

2. AROUND THE WORLD

There are three isolated populations of humpbacks around the world: in the North Atlantic, the North Pacific, and the southern hemisphere (see Fig. 7). Although the waters of the southern hemisphere are continuous with those of the northern hemisphere, humpbacks apparently do not cross from north to south because the seasons in the two hemispheres, and accordingly the animals' breeding cycles, are six months out of phase.

Humpbacks feed in the polar regions and then migrate to the tropics, where they mate and calve. Large concentrations of mating and calving groups are found in tropical waters on each side of the major continents; almost every small island between 30° north and 30° south has had humpback sightings at one time or another. Scattered sightings occur in almost all areas of the oceans, suggesting that some small groups do not follow the general patterns of migration and concentration. Other groups of humpbacks appear to remain in certain areas, notably the Arabian Sea, for the entire year.

The three isolated populations are divided into various smaller stocks that usually return to the same feeding and breeding areas each year. Most scientists believe that there are two stocks each in the North Pacific and the North Atlantic; and perhaps seven in the southern hemisphere. There is some question whether these eleven stocks can be further divided into smaller substocks.

Over 3,000 humpbacks have been tagged in the South Pacific in recent years. Most of the recaptures showed that separate stocks migrate to western Australia, eastern Australia, and islands to the north of New Zealand. According to a study of 40 humpbacks tagged in eastern Australia, 37 were later recaptured there and only three were recaptured in New Zealand waters. Two harpoons lost in humpbacks were recovered more than ten years later in the same waters of Cook Strait, New Zealand.

Tagging studies reveal that humpbacks are segregated into six main areas around the Antarctic feeding grounds and that they tend to return to these areas annually by the same route. Some of the migratory routes overlap, however, and there is evidence of some mixing between stocks. Fifty-three humpbacks tagged off eastern Australia were later caught in the same area where they had been marked years before, but three crossed from eastern to western Australia, and twelve from eastern Australia to New Zealand.

Studies in the northern hemisphere indicate an interchange of animals between Hawaii and Mexico. These

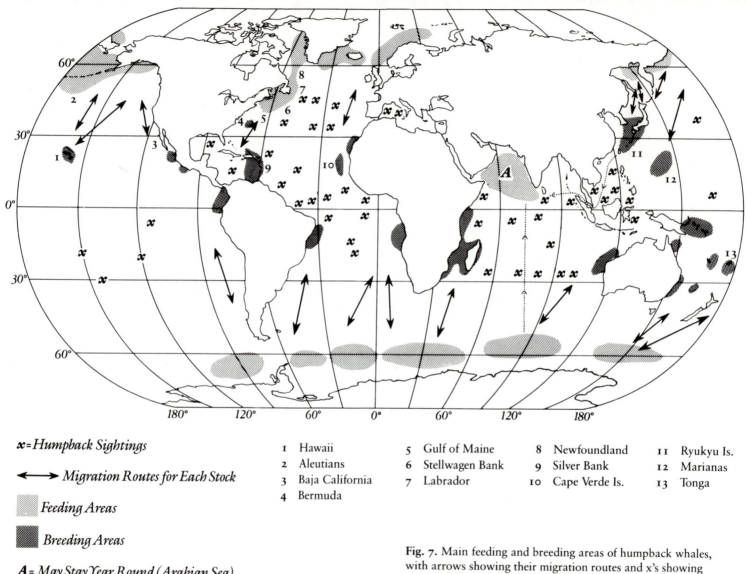

𝐱 = Humpback Sightings

⟵⟶ *Migration Routes for Each Stock*

Feeding Areas

Breeding Areas

A = *May Stay Year Round (Arabian Sea)*

1	Hawaii	5	Gulf of Maine	8	Newfoundland	11	Ryukyu Is.
2	Aleutians	6	Stellwagen Bank	9	Silver Bank	12	Marianas
3	Baja California	7	Labrador	10	Cape Verde Is.	13	Tonga
4	Bermuda						

Fig. 7. Main feeding and breeding areas of humpback whales, with arrows showing their migration routes and x's showing other places where humpbacks have been sighted.
By William C. King.

two substocks intermingle on the northern feeding grounds. The Hawaiian herd or substock appears to be of recent origin, and it may have split off from the Mexican stock. Five animals tagged near the Aleutians were recovered in the Ryukyu Islands of Japan, a 70° longitudinal displacement.

The western North Atlantic stock gives some appearance of being divided into several substocks: one that migrates up the coast to the Gulf of Maine; another that migrates to Newfoundland-Labrador, and a third that migrates to Greenland and Iceland. Comparisons of estimated numbers of animals in various calving areas and feeding areas show similarities that could be explained by the existence of these three substocks, and comparisons of pesticide levels (low in animals that migrate by mid-ocean to Greenland, high in those exposed to the waters of coastal industrial areas) point the same way. But the division remains speculative, in part because the actual migration routes are still poorly known.

Recently Steven Katona and others who study photographs of humpback flukes (tail fins) have been able to show some clear patterns of migration and segregation. Most humpbacks summering off Newfoundland migrate to areas near Puerto Rico, or to Silver Bank, north of Hispaniola, and whales from the Gulf of Maine have also been photographed on Silver Bank. There have been no photographic matches between the Gulf of Maine and Newfoundland-Labrador, suggesting that separate substocks summer in these areas. Individuals found in Greenland and Iceland have also now been shown to go

to the Silver Bank–Puerto Rican area in the winter. It appears, then, that although humpbacks go to the same calving grounds in the winter, they segregate on the summer feeding grounds. To be sure, even if all three herds or substocks go to Silver Bank to mate and calve, they might remain basically segregated from each other while they are there. Obviously we need to know more.

Much less is known about some of the other humpback stocks. In particular, the humpbacks in the Arabian Sea and the northern Indian Ocean are something of a mystery. S. G. Brown, a whale biologist at the National Institute of Oceanography in London, believes that these humpbacks migrate to and from the Antarctic. Yet Dutch biologists have reported sightings during all seasons, with peaks in August–November and January–April, suggesting that animals may come in from the North Pacific as well as from the Antarctic. Still another possibility is that humpbacks remain in this area throughout the year. Not only does the area offer a vast supply of food, but for part of the year its temperatures are satisfactory for calving; indeed, calves have been sighted throughout the year. If this population of humpbacks is indeed non-migratory, it has probably not been hunted for hundreds of years, in which case it could number as high as 500 whales. Hal Whitehead has recently recorded singing in the area at a time when other humpbacks in the southern hemisphere are feeding and in the northern hemisphere are singing, suggesting a separate stock.

Generally speaking, migratory routes are in a north-south direction. Although humpbacks feed in coastal waters and mate and calve in shallow areas around islands or continents, the majority are deep-rooted ocean migrators. Off the west coast of South America and in the western North Atlantic, humpbacks migrate beyond the 100-fathom line. They do not travel in such concentrated groups as when feeding or breeding, but remain more spread out when in deeper water. Sightings of humpbacks have occurred in almost every 10° block of the oceans around the world. When a land mass is in the direct route of the migrators, they may move through coastal waters, as they do when migrating past New Zealand; they may also pass close to land in areas where the 100-fathom line is near shore. But these are exceptions. Thus only a few young strays have been sighted near shore between Florida and Long Island; the vast majority of this stock do not enter coastal waters until they are north of Long Island and onto their feeding grounds.

No apparent relationship exists between migratory paths and oceanic currents and water masses. Humpbacks may travel with or against surface currents, and over every sort of bottom topography. This may indicate that humpbacks use navigational aids such as the sun, the moon, stars, magnetic fields, or sound to find their way in the oceans. The study of the movements of humpbacks is really just beginning. The use of radio tags and photographic identification of individuals will even-

tually help to answer many of our remaining questions about migration and the distribution of stocks.

We will never know how many humpbacks inhabited the oceans before they became prey to humans; one estimate for the southern hemisphere is 102,000. Today the number of humpbacks inhabiting the southern oceans has been reduced to an estimated 3,700. Original estimates for the northern hemisphere are more difficult, since humpbacks in these regions have been hunted for hundreds of years. Today there are about 1,000 humpbacks in the North Pacific and 4,000 or more (perhaps as many as 10,000) in the North Atlantic.

The exploitation of the humpback reached a peak in the early part of this century with the beginning of the modern whaling era. Well over 60,000 humpbacks were killed between 1909 and 1916 in the southern hemisphere, and other peaks of exploitation occurred in the 1930's and 1950's. Over 3,000 humpbacks were killed in the North Pacific during 1962 and 1963. Humpback whaling was banned in the Antarctic in 1939 by the International Whaling Commission but was allowed to continue again in 1949. Finally, after serious depletion of the stocks, it was banned in the entire southern hemisphere in 1963. It was prohibited in the North Atlantic in 1956, and in the North Pacific in 1966. In all areas prohibition occurred only after the stocks were seriously depleted.

The total worldwide population of humpback whales

today is at least 10,000. The world population before hunting was at least 150,000. By far the biggest of the eleven stocks today is that of the western North Atlantic, at least 4,000 whales. There are at least two possible reasons for the comparatively large size of this stock. One is that it has been protected since 1956, whereas humpbacks in the southern hemisphere have been protected only since 1963 and those in the North Pacific only since 1966; yet this time advantage seemingly has not led to any significant increase in the eastern North Atlantic stock. The other reason, which we feel is more plausible, has to do with the particular characteristics of the large calving ground on Silver Bank in the West Indies, where Hal Whitehead estimates that 100 to 300 calves are born each year. Silver Bank has long had a reputation as a very dangerous area. Many ships have been destroyed on its reefs, and sailors and whalers accordingly avoid it. Areas to the south of Silver and Navidad banks still have only a few whales, owing in part to the whale fishery at Bequia in the Grenadines, which, as a so-called native whale fishery, is exempt from the international law.

To estimate an increase in a whale population, the reproductive rate must be determined. For example, using an average yearly increase of 5 percent, with 50 percent females, a whale population of 100 animals would increase to 128 in five years, to 349 in 25 years, and to 1,218 in 50 years. If the original population were 10,000, it would take more than 92 years to achieve that level, because as the population increases, the reproductive rate slows down. Many humpback stocks and substocks around the world have been reduced to 100 animals or fewer. Even if such stocks are now increasing in numbers, it would take many years to detect the increase, since we do not have the ability to detect small changes.

There is no reason to believe that the habitats discussed in this chapter are permanent and unchangeable. By way of example, the Hawaiian humpbacks have apparently been visiting the islands only for the past 132 years; whether their advent was due to harassment in other areas or to climatic changes is not known. Conversely, some established breeding habitats seem to have been abandoned. Dr. N. Miyazaki, of the University of the Ryukyus, reports that there have been no confirmed sightings around Okinawa in recent years, only several unconfirmed sightings by fishermen. There have been no recent confirmed sightings around the northern Marianas. Only one lonely singer was sighted and recorded in the Cape Verde Islands off Africa in 1979, although a mother and her calf were killed there by native whalers the year before.

Having now established in broad terms the worldwide distribution and likely number of humpbacks, we turn next to a more minute examination of their life in the sea, concentrating necessarily on the area of our own greatest expertise, the North Atlantic. Let us begin with the humpback's sojourn in the cold northern waters where it spends its summer feeding.

3. THE NORTH

During the summer months the North Atlantic is tranquil, building strength, it seems, for the first autumn gales. This world of water and sky is vast and intensely lonely. There are no familiar landmarks. The horizon is circular; the ship seems to float in the center of a huge bowl. Water sounds are constant as the sea rushes past the bow, bubbles in the wake.

Once it was thought that the sea contained vast, imperishable resources. Today we are just beginning to understand the delicate balance that exists between creatures that live in the sea, and how little we know about their habits and behavior. For empty as the ocean may seem, it is actually a crowded, noisy place. Below us are canyons and plains as distinct as any found on land. The animals here have their feeding areas, trails, and territories just as their land relatives do. This is the home of the whale, the largest creature that ever lived on earth.

We have come not to kill whales, but to seek information that will help ensure their survival. To most, studying whales seems an adventure; after all, there are cruises to some of the most desirable spots of the world to watch these fascinating and mysterious creatures. And yet, despite the occasional excitement, research cruises are often boring and sometimes unpleasant or even dangerous.

Planning begins months before the actual cruise. Decisions must be made about precise objectives, and routes must be carefully planned to use time at sea to the fullest. Equipment such as cameras, recorders, and hydrophones must be checked carefully: one moment lost because of malfunctioning equipment can mean an important opportunity lost forever. Every moment at sea is precious because of the great amounts of money it takes to support a ship, its crew, and a scientific party. For this reason, no matter what species of whale we have come to study, we will also work with any other species we sight.

Whale watches are set up immediately after leaving port; each member will spend two to four one-hour periods a day watching for whales. Experienced observers are paired with beginners. Sighting whales at sea is not easy; only trained observers can pick out the faint mist of a blow or glimpse a fin among the dark shadows of the sea. After staring into the glaring water for an hour, one can easily come to imagine blows and fins in every wind-tossed wave.

If sighting a whale is hard, identifying its species is harder. Even though each whale species has individual physical characteristics and a distinctive "blow" when it surfaces, it can be difficult to tell one species from another. Often wind will distort a blow when it should be vertical. Moreover, some species exhibit similar behavior: thus sperm, right, and humpback whales display their flukes in the air just before a deep dive. As suggested by Robert Kenney in our laboratory, those species that raise their flukes out of the water tend to be more positively buoyant owing to a higher percentage of fat. Thus they must dive at a steeper angle in order to dive deeply. Only after the experience of several cruises do most observers learn how to interpret what they see and to identify species with only scanty visual clues.

Since whales tend to congregate in the same location year after year, knowing where to look can greatly reduce the time spent searching for animals. Occasionally, the whales find the ship first. Porpoises frequently appear without warning, and like a group of rowdy children surround the ship. Quickly they head for the bow, where the force of the water carries them along like surfers on a wave. Turning on their sides, they look up with one eye. Sounds of whistling and squealing are heard. Impatiently they push each other aside as if vying for the best position. As suddenly as they appear, one moves off, and soon they all depart.

"Blo-o-o-w." The cry from the whale watch brings us running. Hastily cameras are checked. There will not be time to adjust settings or load film; inevitably some of the more spectacular pictures are missed. "Blo-o-o-w." We look along the pointing arm and see a faint mist halfway to the horizon; it is already fading. The ship moves steadily closer. Another blow, and now the whale's back arches to a rounded hump; finally the flukes rise, fluttering in the air for a moment like giant butterfly wings.

From the shape of the blow and the manner of diving, we have identified the whale as a humpback. After a series of breaths at the surface it dives: a deep dive, since the flukes are shown (see Fig. 8). In most instances after showing flukes, a humpback will remain underwater for several minutes.

The ship's engines are shut down. Now we must wait. The silence seems unnatural; only the ship creaks and groans as it wallows in the swell. "Mother Carey's chickens," small petrels, flock around the stern and seem to skip across the ocean surface. Though they appear to be walking on the water with stiff little legs, they are actually flying, touching the water with their feet to steady themselves.

With an explosive whooshing sound the whale surfaces 50 yards away with blows up to 10 feet high. The two fleshy ridges of the blowholes open wide as the animal inhales. Traveling at the surface now, the whale takes a series of breaths before the next long dive. Humpbacks have two blowholes, or nostrils, as do all other baleen whales. (Toothed whales have only a single blowhole.) On very calm days it is sometimes possible to see the exhalation from each blowhole, a "double blow."

There has been controversy over what actually constitutes the visible blow of a whale. Some say it is simply

Fig. 8. Unlike some other whales the humpback shows its flukes on a deep dive. In a shallower dive the flukes do not come out of the water.

the warmer air from the lungs forming a vapor as it meets the cooler outside air. Others say it is a foamy mucus, present in the trachea, which is blown out as the animal exhales. A new, more plausible theory is that when the whale rises a small pool of sea water collects just above the blowhole, trapped by the surrounding fleshy ridges. When the whale exhales forcibly, this water is blown away, forming the spout.

Although Aristotle knew that whales were air-breathing mammals, others ignored this insight for many centuries. Most ancient naturalists and seamen were convinced that whales took in and blew out sea water, and many believed the beasts attacked unsuspecting ships by spouting great geysers of water on them until they sank. This myth persisted, and in 1555 Olas Magnus wrote, "For to the danger of seamen, he will sometimes raise himself above the sail-yards, and cast such floods of water above his head, which he had sucked in, that with a cloud of them he will often sink the strongest ships, or expose mariners to extreme danger." It is not surprising that the blow of the whale gave rise to many myths. To be far from land, in desolate waters, and suddenly have a huge creature appear discharging vapor into the air is a strange and thrilling experience.

Early whalers were aware that whales breathed air, though some believed the spout was poisonous. Herman Melville refers to this belief in *Moby-Dick, or the Whale*: "Wherefore, among whalemen, the spout is deemed poisonous; they try to evade it. I have heard it said, and I do not much doubt it, that if the jet were fairly spouted into your eyes it would blind you." Many people have stood over blowing whales, and none of them, as far as we know, was ever blinded or injured in any way.

Our humpback lingers at the surface preparing for another deep dive, and we are kept busy gathering information. Movies and photographs are taken (see Fig. 9), along with detailed notes on time, water temperature, weather conditions, location, presence or absence of food, and other life in the vicinity. Sound recordings are made on one channel of a tape recorder, while simultaneous comments on the whale's behavior are recorded on a second channel of the same tape. This allows us to determine later the possible correlation between the whale's behavior and the sounds it produces. Depending on our objectives, we may stay with an animal anywhere from less than an hour to over 24 hours. Usually everything goes smoothly, but occasionally there are problems. Sometimes after the first sighting the whale disappears for a whole day or longer; sometimes storms make research impossible and must be waited out. The success or failure of any cruise depends on the whims of nature, and thus is often a matter of luck. Because of these hazards, and since whales offer only brief glimpses into their life cycles, every moment of observation time is crucial.

During the summer months in the western North Atlantic, many humpbacks congregate on shallow offshore areas called banks. These extend from Cape Cod, Massachusetts, to Greenland; they are particularly rich in sea life, and for centuries have been a source of food for both man and whale. During the colonization of Amer-

Fig. 9. Marlin Perkins, aboard a cruise to film an episode of the television show "Wild Kingdom," stands almost directly over a blowing humpback. The double blowholes are fully open as the whale exhales. *Photo by Dave Woodward.*

ica great herds of whales were a common sight on the banks and were often mentioned in logs and journals. In 1594 Robert Dudley made this typical entry while sailing north along the east coast of North America: "After wee weare past the meridian of the Bermudes our courses brought us not far from the cost of Labradore or Nova Francia [Nova Scotia], which wee knew by the great aboundance of whales."

There are frequent references to the great numbers of whales seen off the coasts of eastern North America during the sixteenth and seventeenth centuries, particularly off Cape Cod and Newfoundland. Off Cape Cod twenty years ago we would be lucky to see one or two humpback or fin whales feeding in the spring; once there were hundreds. Although we still always find whales off the coasts of Newfoundland and Nova Scotia, we no longer encounter the vast herds of past centuries.

There is hope, however, for in the last few years many humpbacks have been seen. Our humpback surfaces, its knobby snout parting the water; for a moment one long, white pectoral fin waves lazily in the air. Of all the large whales the humpback is perhaps the strangest-looking with its knobs, warts, and barnacles.

Tales of sea monsters may well have originated from a first meeting with a humpback whale. Thus in July 1734 the Rev. Hans Egede, the "apostle of Greenland," encountered a fearsome animal in the North Atlantic that could have been a humpback. In his words, "There appeared a very large and frightful sea monster, which raised itself so high out of the water that its head reached above our main-top." (Humpbacks are notorious for their great leaps out of the water, though in this case the frightened observer surely exaggerated.) "It had a long sharp snout, and spouted water like a whale, and very broad flappers. The body seemed covered with scales and the skin was uneven and wrinkled, and the lower part was formed like a snake." (All these observations suggest the humpback but the last, which we may attribute to a vivid imagination.) "After some time the creature plunged backwards into the water, and then turned its tail up above the surface a whole ship length from the head." (This could describe a humpback performing a deep dive and showing its flukes.) "The following evening we had very bad weather."

As recently as June 1877 a Lieutenant Hayse, aboard the Royal Yacht *Osborn*, saw a head and two "flappers" about fifteen feet long and concluded that he had seen a sea monster. We know that the humpback has flippers about fifteen feet long and sometimes lies at the surface and paddles with them (see Fig. 10).

In the last several hours our humpback has been moving continuously. We follow until dark, and the humpback continues on its way. Since most of the whale's time in northern waters is spent feeding, its wanderings may be directed by the movements of its food. Without great amounts of food, the humpback cannot accumulate enough blubber to sustain it throughout the winter months, when feeding does not take place.

Fig. 10. Aerial photograph of two humpback whales. Note the visibility of the long white pectoral fins.

Wm C. KING

4. FOOD AND FEEDING

Because of the whale's great size it must be supported by vast amounts of food. During the summer humpbacks not only must consume enough food to stay alive, but must build up reserves of fat for wintering over on the mating and calving grounds, where they do not feed. The average-sized humpback needs about one ton of food per day, an intake of a million calories or more.

The summer feeding grounds are quite different in the northern and southern hemispheres. In the North Atlantic, humpbacks feed off the coasts of the United States, Canada, Greenland, Iceland, and Norway. In the eastern North Pacific, humpbacks summer off Alaska and the Aleutian Islands to the Bering Sea. The western North Pacific stock feeds from the Sea of Okhotsk to the Bering Sea.

The diet of humpbacks in the northern seas is varied; they consume fish, crustaceans, and sometimes pelagic mollusks such as squid and pteropods. They feed near areas long inhabited by humans, and in fact they were one of the first cetaceans to be hunted themselves. As coastlines became more populated, fishermen and seamen increasingly encroached upon the whale's territory. More and more the humpback is in direct competition with humans for food.

Humpback whales in the South Pacific, South Atlantic, and southern Indian oceans feed in areas off the barren coasts of Antarctica. Here the cold, fresh Antarctic waters meet the warmer, saltier waters of the southern oceans; the meeting creates an area of upwelling, the Antarctic convergence, an area rich in sea life. Just south of the Antarctic convergence, where the water is colder, the humpback feeds almost exclusively on shrimp-like crustaceans called krill, which occur here in great swarms. Because the land is uninhabited and whale hunting is no longer allowed, the humpback can feed unmolested in Antarctic waters.

Among the items in a humpback's diet are herring, capelin, smelt, cod, sand lance, pollack, sardines, salmon, and anchovies. Six cormorants were found in the stomach of a dead humpback washed up on the coast near Berwick, England, in September 1829, and another was lodged in its throat. It was presumed the whale had choked in an attempt to swallow this last bird. Since birds frequently flock around feeding whales (see Fig. 11), even snatching food from their mouths, it is likely that the cormorants were swallowed accidentally by the whale while it was feeding. Millais observed the behavior of birds among feeding whales:

Fig. 11. A group of humpbacks and birds in a feeding frenzy off the coast of Alaska. *Photo by Charles Jurasz.*

While feeding is in progress, quantities of the crustaceans drift to the surface and are attacked by the pelagic birds, such as the kittewake, the greater and dusky shearwater, and Wilson and Leach's petrels. I have seen as many as a hundred Leach's petrels swarming on one spot no larger than two square yards and picking up the "bait" as it is called by the whalers. Most of the birds keep their wings extended the whole time, and fly off to assemble again in an instant as soon as another whale's back shows.

Birds are not the only animals to share the humpback's dinner. We have often seen white-sided dolphins swim beside, over, and in front of the heads of feeding humpbacks. The more active the whale, the more active the dolphins, which suggests that the dolphins feed on fish not caught by the whale.

When humpbacks were hunted their stomach contents were often examined and measured, and from the resulting data we can get some idea of the capacity of the whale. A 45-foot female humpback's stomach contained 600 large herring. From two male humpbacks, 32 and 35 feet long, 800 walleye pollack and 1,200 pounds of saffron cod were taken. Another 46-foot female had consumed 1,000 pounds of plankton.

The Abbé Bonnaterre, a French biologist, wrote in 1789:

When the Jubarte [humpback] wants to eat, it opens its mouth wide and swallows a lot of water with its prey; then you see the folds of the belly open very wide. In this moment the contrast of the pretty red that shines in the cavities of the folds along with the black of the baleen that are attached to the jaws and the white shining from the mouth, all goes to produce a very agreeable effect.

In the early nineteenth century, the French zoologist R. Lesson wrote of the southern hemisphere humpback:

The whale does not so much seek its food as its food seeks it. The sea is often very rough, and the height and violence of the waves is such, that the spray in breaking over the vessel brings along with it great quantities of medusa and flying fish. It is under these circumstances that these giants are most active, as if enjoying the storm, and then appear busiest in pursuing their prey.

The naturalist B. Rawitz, another author of this period, believed that in order to close its mouth the humpback had to throw itself upon its back.

Others observed feeding behavior with greater accuracy. In 1632 Théodat Sagard observed feeding humpbacks in the Bay of Gaspé that were evidently striking the water with their tails or fins, making a great deal of noise. He was told that this was to surprise and mass together the fish, which were then swallowed by the whale. Ingebrigtsen published a paper in 1929 on this same feeding behavior, describing two methods used by the humpback in taking krill. One was to lie on its side on the surface and swim round in a circle at a great rate, lashing the sea into foam with strokes of its tail; this forced the krill toward the center of the circle, where they were conveniently devoured by the whale. In the

Fig. 12. In this group of feeding humpbacks three have surfaced with their mouths open, showing the baleen in the upper jaws.

Fig. 13. This humpback is lunge feeding. Note the bagged throat distended from taking in quantities of water and food.

Fig. 14. A feeding humpback surfaces, straining water through its baleen plates and leaving the food behind in its mouth. *Photo by John Arsenault.*

second method the whale circled below the surface continuously blowing off air, which rose as a thick wall of bubbles, driving the krill into the center of the circle. Recently these and other methods of feeding have been observed by Charles and Virginia Jurasz in Alaska, and by researchers at the University of Rhode Island in the North Atlantic.

The behavior of a feeding humpback depends on the kinds and quantities of food it is pursuing. When food is abundant the humpback uses a technique called lunge feeding. The whale may approach the food from under-

neath or from the side, opening its mouth just before surfacing and lunging through the food as it takes in both food and water (see Fig. 12). As it lunges, the ventral grooves under the chin expand and form a large pouch (see Fig. 13); the pleated skin allows the humpback to take in much larger quantities of water and food than would otherwise be possible. Then the whale forces out the water through the baleen plates with its tongue, leaving the food behind to be swallowed (see Fig. 14). It was once generally accepted that humpbacks swam along the surface with their mouths open, skimming food from the top of the water. But T. Nemoto, a Japanese biologist, has found that humpbacks never skim food at the surface as the blue and bowhead whales do, nor do they take scattered plankton.

Humpbacks sometimes feed on fish by making rapid, sudden snatching movements with distinct half turns around the body axis. As we have seen, in the northern hemisphere, where humpbacks feed more frequently on fish, a greater percentage of whales have white pectoral fins. Paul Brodie, a Canadian scientist, has speculated that these white pectoral fins may be an adaptation that aids the humpback in capturing fish. When the humpback approaches a school of fish holding its fins out horizontally, the fish see two white bars with a dark area, the head, in the center. Seeking escape, the fish swim toward the dark area and are trapped. In the southern hemisphere the pectoral fins of humpbacks are darker, perhaps because these whales rarely feed on fish.

The humpback has several methods of bubble feeding. One is to release air underwater, which results in a cloud made up of many small bubbles some five to eight feet in diameter. As the cloud rises, it expands at the surface to 20–25 feet, serving to camouflage the whale as it surfaces and swims through the bubbles to feed (see Fig. 15).

A bubble net is formed when the humpback swims underwater in a circular pattern, either horizontally or in an upward spiral, and releases bubble columns at regular intervals. The bubble columns form a circle at the surface. Sylvia Earle, a marine biologist, has found more krill within the bubble net than in areas surrounding it, evidence that the wall of bubbles either concentrates or contains prey within the circle. When the circle is almost completed, the whale turns to the center and opens its mouth to feed (see Fig. 16). The size of the bubble net can vary from about 30 to 100 feet. Bubble nets can also be U- or V-shaped, with the whale approaching the food from the open end of the figure.

Unlike some of the other baleen whales, the humpback feeds both near the bottom and at the surface of the ocean. A humpback feeding near the bottom performs steep dives, displaying its tail flukes vertically in the air. It is not clear whether the whale habitually feeds off the bottom or does so only when it is forced to chase fish into deeper water. For whatever reason, humpbacks apparently go to great depths to obtain food, as the following story illustrates.

In November 1904 the submarine cable in Alaskan waters between Valdez on Prince William Sound and Sitka was suddenly interrupted. The cable had been laid only a month before and had been in perfect working order. Tests were made, the trouble was located about ten miles from Sitka, and in January a cable ship set out to make repairs. As the crew began to raise the cable, they noticed that there was a considerable strain on it. Thinking it was caught under a rock, they maneuvered the ship to loosen it. Slowly the cable was pulled aboard, until suddenly the carcass of a whale appeared. A loop

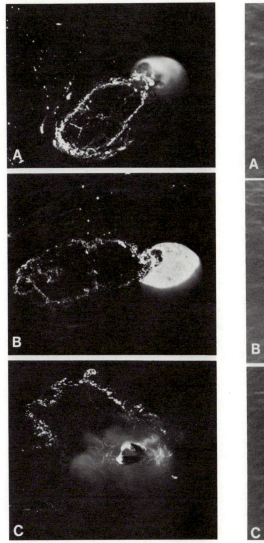

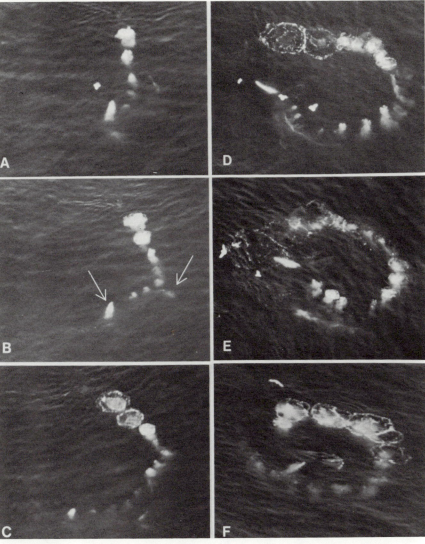

Fig. 15. A humpback using the cloud net method of feeding. In A the humpback releases air underwater, producing a five-foot bubble. The bubble comes to the surface (B), followed by the whale surfacing with its mouth open (C). *Photo by Ann Brayton.*

Fig. 16. Sequence showing a humpback blowing a bubble net to capture food. In B arrows show the white fin and flukes. When the bubble net is completed in F, the humpback surfaces in the center with its mouth open scooping up its entrapped prey.

of cable was twisted and wrapped securely around the lower jaw of a 50-foot humpback whale!

The humpback had evidently been feeding, swimming slowly along the bottom when its open mouth encountered a part of the cable that was suspended on bottom irregularities. In the effort to get free, the whale looped the cable over its jaw and became trapped. Struggling, the animal had twisted and torn the cable, making several breaks in the conductor and rendering it inoperative. The depth of water where the humpback became entrapped was 65 fathoms. Recent studies have shown that humpbacks forage as deep as 100 fathoms.

Individual feeding styles were observed over a long period by Stormy Mayo and Carole Carlson at Stellwagen Bank off Cape Cod. One animal which they named Binoc would breach, its body coming one-third or more out of the water, fall back, and emerge in a horizontal lunge with its mouth open. Another which they named Cat's Paw would come up vertically, climbing halfway out of the water with its mouth open, submerge, then come up again horizontally with its mouth open so that only its head was out of the water. One named Speckles would produce a cloud of bubbles, surface through it, thrash its tail, dive, then lunge upward in a horizontal position with its mouth open. A few whales were observed blowing bubble nets consistently. One wonders if these feeding styles have varying efficiencies.

The humpback has learned to take advantage of man's invasion of its feeding grounds. Ignoring men, nets, and boats, the whales will gather around fishing boats and devour the leftovers. Estimates have been made that fish discarded from trawlers alone may total 200,000 tons per year. David Sergeant, a Canadian biologist, notes that "this statistic does not measure the undersized fish escaping from the trawls, which is what the whales presumably take, but gives an idea of the quantities of fish that may be available to the whales."

In 1859 A. E. Verrill observed schools of humpbacks and fin whales in the Bay of Fundy feeding on shrimp and herring that were being seined:

The whales were feeding both on the herring and shrimp, and were so tame and so intent on their feeding that they often came within an oarlength of the numerous boats and vessels engaged in seining the herring, often, indeed, passing under the bow-sprit of the vessels. At that time they were never disturbed by the fishermen, and they rarely came in contact with the nets and boats, which they carefully avoided by turning aside or diving under them. There were dozens of them in sight at once. Many that I saw were 60 to 75 feet long, often exceeding the length of the schooners, alongside of which they often passed near enough to be touched with an oar. It was a rare and imposing sight, never to be forgotten, to see these leviathans so tame and fearless of man. One large hump-back whale, which was easily recognized by means of a large barnacle attached by the side of the blow-hole, so as to cause an abnormal noise in blowing, had frequented these waters every summer, for more than twenty years, according to the fishermen. At that time there were more than 50 vessels fishing at this place, each with 4 to 6 boats and seines in use.

More recently D. Sergeant reported observing a humpback on the Grand Banks off Newfoundland feeding with the most exaggerated leaps and dives, totally indifferent to three vessels that crisscrossed its position dragging for fish. At sea the whale and the fisherman may be mutually tolerant, but off the shores of Newfoundland the humpback is regarded as an enemy.

In the 1970's whales became entangled with increasing frequency in fixed inshore cod traps in Newfoundland. The trap fishing season there lasts for almost three months, from June to August. Since one cod trap can yield the fisherman $5,000 worth of fish in a single day, the destruction of a single net, valued at $8,000, can represent a sizable loss; the loss in 1979 alone was well over $1,000,000. Between 47 and 57 of the 71 to 90 whales entangled in traps that year were humpbacks (the rest were Minke and fin whales), and 10 to 12 of these died in the nets. Peter Beamish and Judy Perkins have identified the problem: the entrapments invariably take place at the time capelin come inshore to spawn on the beaches, and occur because cod and whales follow the capelin inshore. Some of the entrapped whales become entangled in the nets and drown, others may die later from injuries, and still others escape unharmed.

This problem is not limited to the western North Atlantic, but occurs also in the western Pacific. In Australia in 1974 and 1977 humpbacks became entangled in meshes set to protect surfers from sharks, but were released alive. Beamish feels that with our present technology this problem could be alleviated within three to four years by using a sound warning system that would ward off the whales without disturbing the fish. John Lien, a psychologist, has recently had some success keeping whales out of nets by attaching underwater bells to the meshes.

In 1976 we spoke to an old Newfoundland fisherman who assured us that this problem is a recent one. In all his years of fishing he had never caught a whale in his nets, although he could recall hearing about such an incident many years ago. There may be several reasons for the recent increase of entrapments. The humpback population may be increasing, or humpbacks that previously fed offshore may now be moving inshore because the capelin in deeper waters have been seriously overfished in recent years. There has also been a great increase in the number of cod traps being set. Whatever the causes, it is hoped that some solution will be found that benefits both the whales and the fishing industry.

In the last few years things have improved, according to John Lien and Hal Whitehead. Interested persons and fishermen are quicker to free humpbacks from the nets, thus reducing their mortality, and fewer humpbacks have come inshore since the capelin stocks have recovered off Labrador. It is clear that humpbacks go to their traditional feeding grounds and if for some reason their food is at a low level, they search out more concentrated sources of supply. Off Cape Cod humpbacks fed on herring fifteen or so years ago. The herring populations

Fig. 17. A baleen plate from the upper jaw of a humpback whale. *Photo courtesy of Smithsonian Institution.*

were soon depleted and were replaced in the humpback's diet by sand lances, which have a different distribution that has now become the distribution of the humpback (also the fin whale). Since the sand lance is found closer to land (e.g. off Provincetown), this results in excellent offshore whale-watching.

As the whale population increases, entanglements in nets will become more serious. As we were putting the final touches on this manuscript in October 1984, a humpback female named Ibis was potentially in serious trouble. She was first seen swimming around with a gill net entangled in her baleen on October 6. Rescue attempts at first were unsuccessful, but on November 22, Thanksgiving Day, Dr. Mayo and his associates from the Provincetown Cetacean Research Program were able to remove the gill net. Later, on December 9, Ibis was seen in the vicinity of 30 other humpbacks and appeared to behave normally.

No matter what the humpback is feeding on, it relies on its baleen to strain the food from the water. The upper jaw is lined with these thin plates of bone, and an adult whale has from 270 to 400 plates on each side of the jaw. Each plate is fringed with coarse, gray-brown bristles similar to broom bristles (see Figs. 17, 18). Baleen plates are present even before birth, measuring about one and one-half inches in a near-term fetus. Humpbacks in the northern hemisphere have shorter baleen (33.5 inches is the longest recorded) than those in the southern hemisphere, where the baleen averages 43 inches. Perhaps longer baleen is necessary for the efficient capture of Antarctic krill.

Whalers referred to the baleen plates as "whalebone,"

Fig. 18. A front view of feeding humpbacks with their mouths open. The bottom jaws and throats are distended with a large quantity of food and water; the upper jaws show the hairlike baleen.

once a valuable product that served many purposes. It was cut into strips for use in women's corsets, and fashioned into hoops for crinoline skirts and umbrellas. Woven strips were used for chair seats, bed bottoms, carriage sides, and window guards. Shredded whalebone was made into nets and brushes, and added to wigs to help keep the shape of the curls. The baleen of humpback whales, being coarse and dark, was less highly valued than that of the right whale.

How does a humpback locate food? A humpback's vision is about as good as a human's underwater. It is said that whales do not have very good binocular vision because the eyes are located on the sides of the head; but in humpbacks, at least, both eyes bulge out and can be seen at the same time by someone facing the animal, thus arguably affording it much better visual perception than it was once thought to have. Vision must be important to humpbacks in finding food, even though the sea is often murky even in daylight.

The humpback has another potential sensory organ, its sinus hairs or vibrissae. These are single hairs that protrude from each tubercle, or bump, located on the whale's upper and lower jaws (see Fig. 19). At one time vibrissae were considered to be remnants of a hair covering and to have no functional significance. Investigators have found, however, that each vibrissa is attached to a clearly visible nerve cord and has a well-developed blood supply, suggesting that these hairs function as a sensory organ. Toothed whales do not have vibrissae, and among baleen whales the humpback has the thicker.

The theory is that the humpback uses its vibrissae much as a cat uses its whiskers to feel the presence of close objects. Since the vibrissae are sensitive to the touch of foreign objects, the humpback can use them to determine the immediate presence of a food source. If large concentrations of food are present, the vibrissae tell the humpback when to open its mouth to obtain the greatest amounts. Paul Brodie theorizes that the humpback is more reliant on tactile feeding because it does not move as fast as other rorquals.

Feeding humpbacks, whether dining on fish or krill, all seem to have one characteristic in common: bad breath! Dr. J. Forster, who returned to England with Captain James Cook, wrote of encountering humpbacks near Tierra del Fuego: "When between Tierra del Fuego and Staten Island, Lieutenant Pickersgill was sent to Success Bay, and on this occasion it was remarked that no less than thirty large whales played about them in the water. Whenever they were seen blowing to the windward, the whole ship was infected with a most detestable rank and poisonous stench, which went off in the space of two or three minutes."

J. G. Millais stated that earlier scientists such as W. Lilljeborg, C. Jovan, and E. G. Racovitza all bear testimony to the warm and fetid breath of the humpback whale; and he himself says that "all large whales are foul

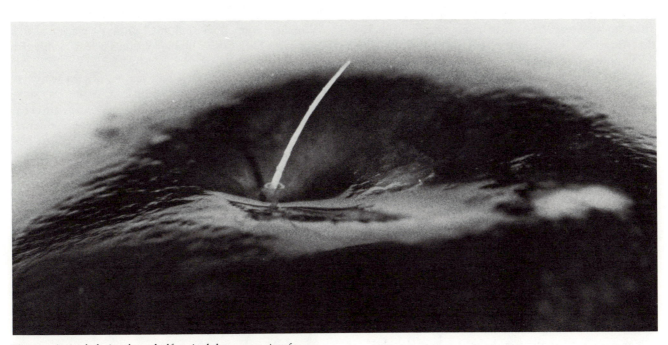

Fig. 19. A single hair, about half an inch long, growing from a tubercle on the lower jaw of a humpback whale. *Photo by Geoffrey LeBaron.*

in this respect, the humpback particularly so." Raco-vitza stood several times over a spouting humpback in the Antarctic and reported that the breath of these whales possessed a nauseating odor. William Schevill, a cetologist from Woods Hole Oceanographic Institution in Massachusetts, also attests to the unpleasant breath of western North Atlantic humpbacks. But we have stood over blowing whales in the West Indies and have never been aware of any odor. Evidently bad breath is a condition of feeding whales; since the humpbacks in the tropics are not feeding, their breath has no odor.

Much of what a whale eats is converted to fat and stored. After a summer of feeding, the blubber layer of the humpback is at its maximum thickness (up to 7.5 inches). The whale must live off this fat throughout the winter in the tropics. Humpbacks have thicker blubber relative to their size than any other rorqual, and in absolute thickness of blubber are second only to the blue whale. Pregnant females are particularly fat, with up to seven tons of blubber. Some young humpbacks have thicker blubber than adult whales.

The seasonal variation in the thickness of the blubber layer probably accounts for whalers' varying opinions of the humpback's worth in oil, which was used for illumination and in soaps. Many whalers considered the humpback to be poor in oil and not worth going after. The English naturalist Richard Lydekker took this opinion as a fact and offered an odd comment on it: since the humpback was easy to capture, he wrote, "it must have an intuitive knowledge of the poorness of its oil and the shortness of its bone [baleen]." Others have stated that the humpback's thick blubber renders it remarkably rich in oil for its size.

At the end of the breeding season, after a winter of fasting, a whale's oil production is at its lowest. Some whalers were aware of this, and Collins wrote in 1892: "Whales when migrating north are poor, but on their return south are invariably fat and contain about fifty percent more oil than when on their northern passage." Charles Scammon, a Californian whaler who discovered the breeding grounds of the gray and humpback whales off the coast of Baja California, also found that the humpback varied more than other whales in the production of oil.

While the humpbacks are on the feeding grounds, they are usually found singly or in groups of from two to five. Not much is known about their daily movements while on the feeding grounds, but some are probably related to the movements of the species they are feeding on. William Schevill and Richard Backus spent several days in October and November 1958 observing a humpback off the coast of Maine that swam through the same area at the same time each day. The time did not alter with changes in the tide, as it would have if the location of the humpback's prey varied with the tides. They could not tell whether or not the whale was feeding, but "whatever it was doing, it evidently followed the sun and not the moon."

Other researchers have noted that feeding activity seems to be higher during the early morning and late afternoon hours, suggesting a correlation with the diurnal migration of food species. Jeff Goodyear, using a radio tag, has obtained evidence that humpbacks also feed at night. All in all, we still have a great deal to learn about the feeding habits of humpbacks.

As the summer ends, the time comes to group together in preparation for the move southward. For about four months the whales will go without any substantial amounts of food. No one knows exactly what triggers this migration. Perhaps the pregnant females sense that their time is near; perhaps the whales sense differences in daylight length or water temperature. In any event, the long journey begins.

5. JOURNEY SOUTH

Humpback whales migrate with great regularity, and since they feed, mate, and calve close to land, more is known about their habits and migrations than about those of most other whales. In the western North Atlantic some humpbacks begin to move south in October. The first to leave are the mature females accompanied by their yearling calves. These calves will be weaned soon, and when they return to the north in the spring they will feed on their own. Next come the immature males and females, followed by the resting females (females not ovulating) and the mature males. The pregnant females bring up the rear. Males are most abundant in the center and rear, but are found in all parts of the herd.

During the summer feeding months humpbacks are found singly or in groups of two to five. As the migration begins, groups join together, forming herds. A century ago, before modern whaling took its toll, ships often encountered great herds of migrating humpbacks, some numbering over a hundred animals. Today we rarely see herds this large.

The humpbacks do not all leave at the same time. Animals further to the north may start later, so that the migration actually occurs over a couple of months, with some leaving in October, some in November, and some in December. The timing of some groups' departure is remarkably regular: thus humpbacks invariably leave the Stellwagen feeding banks off Cape Cod between November 14 and November 26. Their journey is long, up to 4,000 miles, and it will take the whales a month or two to complete it. Traditional routes are followed, and generally the time they pass a given point is predictable, varying only a few days from year to year. Sometimes there is a greater variation, which may be due to extreme yearly climatic variations, especially in temperature.

No one really knows how whales manage to find their way each year, but they travel as if on a well-marked road. Some scientists believe they may use low-frequency sound waves, which bounce off familiar bottom topography, or sense temperature differences in the water. They may also use visual cues, such as orienting by land masses or even by the sun during the day and the stars at night.

Researchers have discovered that birds can orient themselves by sensing the earth's magnetic field; perhaps whales also have this ability. G. B. A. Bauer and his colleagues recently found a magnetic material, probably magnetite, in humpbacks, which in theory could enable them to detect the earth's magnetic field. Joseph Kirschvink believes that whales strand more often at locations along the coast with low magnetic intensities, and M. Klinowska has come to similar conclusions. The use of magnetic fields along with other cues would explain the precision of humpback migrations over thousands of miles.

Little was known about the distribution and migrations of baleen whales until offshore whaling began. The first indication that rorquals migrated came in the late nineteenth century, when some blue whales caught at a Norwegian whaling station were found to carry fragments of American bomb lances. In 1920 a Norwegian biologist, J. Hjort, began to mark whales systematically with a copper lance, but for some reason his tags did not remain implanted in the blubber and were lost. The Discovery Committee of the National Institute of Oceanography, in London, subsequently developed an effective tag consisting of a stainless steel tube about 10½ inches long, which was fired from a harpoon gun. These tags usually did no damage to the whale, and did not deteriorate even after being lodged in a whale's blubber for 25 years.

Many of these tags were recovered on the whaleships while the whales were being processed; those overlooked during the processing were usually found later in the boilers along with the oil. Although stamped instructions on the tag promised a reward to the finder if it was returned to the National Institute of Oceanography, few rewards were claimed. Of the many thousands of tags placed on whales, only a small fraction were ever recovered by the Institute.

We might also mention an amusing case of accidental tagging. In June 1954 a can of tooth powder was discovered in the stomach of a humpback whale killed off the coast of New Zealand; inside it was a piece of paper with the name and address of a crew member of the whaleship *Willem Barendsz*. The man had thrown the can overboard in January 1954, during the whaling season, and the whale probably swallowed it while feeding.

During the period when humpbacks were hunted, recovered tags provided a great deal of information about the whales' migrations, seasonal movements, distribution, and age. Biologists also had the opportunity to examine dead whales and gather data on their internal organs and physical characteristics. During the 1950's some of the greatest contributions to our knowledge of the humpback's migrations and biology were made by R. G. Chittleborough of Australia and W. H. Dawbin of New Zealand. Since the humpback can no longer be hunted, their studies may never be duplicated.

With the cessation of whaling, new tagging methods had to be devised to study whales' movements. One such method is a visual tag featuring brightly colored streamers or the like; unfortunately these tags do not remain on the whale very long. Another method is a radio tag containing a small transmitter. The transmitter emits a signal that can be tracked by researchers on a boat or airplane; some signals can even be tracked by satellite.

An ingenious radio tag has recently been developed that consists of a ⅝-inch cylinder containing a transmitter and equipped with an eighteen-inch whiplike antenna. Since it weighs only a pound, it can easily be fired from a modified harpoon gun. The tag penetrates about ten inches into the whale's blubber and muscle, and is comparable to an injection with a hypodermic needle; its tip is coated with an antibiotic to prevent infection. If the tag is placed high on the whale's back, the signal range can be as far as 50 miles. The transmitter turns off when the whale is underwater, saving the life of the battery, which lasts from two to eighteen months. Eventually the tag falls off the whale.

By this method a humpback was successfully tracked by a team from the National Marine Fisheries Service headed by Michael Tillman. They were able to track the whale for six days for a total distance of fifteen miles in Glacier Bay, Alaska. The tag was still operating when the scientists had to return to shore because their boat charter had expired. The next humpback tagged was tracked for seventeen days; it stayed in the bay the entire period. A gray whale has now been tracked 2,800 miles for 95 days. More recently Jeff Goodyear has successfully tested his "Remora" radio tag, which attaches to the whale with a suction cup.

As this was written, a humpback with a radio tag was for the first time tracked by satellite, by Bruce Mate. It was tagged off Newfoundland and for six days moved offshore over the Grand Banks at about three and a half knots. According to Peter Beamish and S. Carrol, the sounds of this whale as monitored before and after the tag attachment showed less change than the sound responses of other humpbacks to boat and diver movements. Soon whales will be tracked by satellite for long periods, giving researchers a better understanding of seasonal movements and of migration routes.

Radio and visual tags are useful only for short-term surveillance of whales. When we began our work with humpback whales, we attempted like most other researchers to photograph each animal we encountered. Humpbacks have distinctive patterns of barnacle growth, pigmented skin on their flukes, varying dorsal fin shapes, and various other scars and scratches; we hoped that these physical differences would allow us to identify individual whales. The next step was for someone to gather as many photographs as possible and make a systematic attempt to distinguish individuals.

A few years ago Steven Katona and Scott Kraus, from the College of the Atlantic in Maine, initiated such a program. First they collected photographs of humpback whales in the western North Atlantic from scientists and others who had worked with them. When they studied these photographs, they found that the pattern on the underside of the flukes seemed to be the most promising means of individual recognition. Although the top of the fluke is usually dark, the underside ranges from almost pure white to almost pure black, with many various patterns of black and white.

Katona found that each humpback has its own unique color pattern on the fluke (see Fig. 20). After examining 78 photographs of different humpback whales, he determined that each could be individually identified: in effect, the humpback possesses a natural tag. By matching the photographs of whales with identical patterns on their flukes, Katona could show the movements of individual whales. One humpback photographed at Bermuda on April 15, 1976, was later photographed off the coast of Maine in July 1976. Photographs taken on October 1, 1974, off Nova Scotia and in July 1976 off Maine were demonstrably of the same humpback; in this case it was shown that the pattern remains unchanged for at least two years. One animal photographed alive in July 1976 was found dead a month later. Although the carcass was in poor condition when found, the fluke pattern was still discernible.

The potential for this method of individual recognition is enormous. Unlike tagging, it poses no threat to the animal. There is not a great deal of expense involved, and since the animal naturally displays the tail flukes before a dive, photographs are easily obtained. Best of all, since almost every individual has been differently marked by nature, the potential sample size is much greater than it would be with conventional tagging methods.

The success of this project depends on the cooperation of all the scientists, photographers, and others working with the humpback whale in the North Atlantic. Katona has already compiled two photographic catalogs of individuals and over 3,000 individuals have been identified (see Fig. 21). Researchers in Hawaii and Alaska are also developing a catalog of individuals. Eventually a great deal may be learned about movements, social organization, stocks, age, population size, and migrations of the humpbacks in all the oceans. Other significant findings will be discussed in appropriate chapters.

By December the first humpbacks migrating southward have arrived in the West Indies. Each group heads for its traditional mating and breeding ground. The western North Atlantic stock of humpbacks distribute themselves from Grand Turk Island, north of Hispaniola, to Venezuela. Humpbacks in the West Indies have somewhat specific requirements for their calving and mating areas. They choose banks, usually around islands, with a water depth of from 10 to 100 fathoms. These banks must be over two or three miles wide, which is why many narrow banks along the islands, particularly on the Caribbean side, are unoccupied. A few animals are found in deep water around pinnacles, such as in Mona Passage. Mothers about to calve and those with calves usually go into sheltered bays or stay in the

Fig. 20. The flukes or tail fins of humpback whales show individual variation in color pattern; as with fingerprints, no two are exactly the same. *Photo courtesy of Charles Mayo.*

shallower waters around exposed reefs, presumably for protection, especially when the calf is very young. Others occasionally go into one or two fathoms of water on sloping beaches.

In the West Indies most humpbacks are found on three banks extending southeast from the Bahamian Archipelago: the Mouchoir, Silver, and Navidad banks. Mouchoir Bank (Handkerchief Shoals) was named for its shape; because it is shallow and has many exposed reefs, this dreaded bank has always been approached cautiously by ships. Its 372-square-mile area is only occasionally visited by humpbacks. Navidad, the bank closest to the Dominican Republic, is the deepest of the three, with no exposed reefs. Navidad Bank is some 204 square miles in area; usually about a hundred humpbacks are to be found there in breeding season.

Situated between the other two banks, Silver Bank is the largest in area. Its 700 square miles serve as a breeding and calving ground for some 800 or more humpbacks (perhaps well over 3,000), 75 percent of the western North Atlantic stock. Because of a fringing reef on the northeast edge, Silver Bank was the greatest peril to Spanish mariners sailing home from Hispaniola; it probably contains more sunken ships than any other spot on earth. Some speculate that $600 billion in gold, silver, and other Spanish treasure from the New World lies buried on Silver Bank.

Some treasure has been recovered. Although Sir William Phipps, the first Royal Governor of Massachusetts (in 1692), ended the persecution of persons believed to be witches, he is probably better known for his recovery of treasure. In 1683, Phipps, then the captain of a trading sloop from Massachusetts, approached King Charles II with a plan to salvage the treasure on Silver Bank. This voyage was unsuccessful, but while in Puerto Plata, Hispaniola, Phipps became acquainted with an old Spaniard who told him where to find the wreck of a galleon. He returned to England to seek aid, but the King refused. He finally obtained assistance from Christopher Monch, Second Duke of Albemarle. Two ships, the *James and Mary* and the *Henry of London*, were purchased, and Phipps set sail for the Caribbean.

In January 1697 Phipps's second mate, William Covell, and Francis Rogers located the treasure on Silver Bank while Phipps was on shore trading with the Spaniards. Eventually Phipps and his men recovered more than $600,000 worth of silver and gold, plus jewels and other articles. Later others recovered still more treasure.

The story interested us not because of the treasure but because Phipps's search took place in January, a time when Silver Bank today is well populated with humpback whales. We thought that by this early date some of the whales might have moved to this relatively isolated area, as the human population grew and shipping activity increased in the Caribbean islands. If humpbacks were there, we thought Phipps would surely have mentioned them in his log book or journal.

A portion of a transcript entitled *Some brief remarks*

Fig. 21. Physical anomalies can also be used to identify individuals. The humpbacks shown here, one with half of its fluke missing and the other with an unusually shaped dorsal fin, could be quickly identified. The whale with the missing half-fluke was named Silver in 1978 by Stormy Mayo and Carole Carlson. The fluke was that way, with all scars completely healed, in 1978. She also had deep scars on her back. She had a calf named Beltane in 1980 who has returned to Stellwagen Bank each year since. She had a second calf named Aster in 1983. *Fluke photo courtesy of Charles Mayo.*

upon a voyage made by the James and Mary and Henry of London for the Banks of Bahama in America, written in 1686 or 1687, was sent to us by Katherine Osborne, a researcher for a group interested in Phipps's treasure. In this transcript was the following comment: "Those parts abound in spermacety Whales which often times appear like those Rocks in ye water."

As we have seen, it is easy enough to confuse humpbacks with sperm whales. F. D. Bennett, in his book *Whaling Voyages* (1840), says of the humpback, "When seen on the surface of the water it bears a close resemblance to the Sperm Whale in color and the appearance of its hump, as well as in the habit it has of casting its tail vertically in the air when about to dive." Since sperm whales do not congregate in shallow water, the evidence suggests that humpback whales have been using Silver Bank as a mating and calving area for at least three hundred years.

In 1977 treasure hunters again visited Silver Bank. For a while it seemed as if the humpback would become the victim of a dispute over the ownership of the bank, now claimed by both the Dominican Republic and Grand Turk islands. Each government, claiming jurisdiction over Silver Bank, granted permission for a group to search the bank for treasure. Rumor had it that the ship from the Dominican Republic had dynamite aboard and was planning to use it to break up the coral deposits that might be covering the wrecks. Since it was the mating season, some feared that the use of explosives would disturb the whales, but the stories were quickly denied and no blasting took place. At the same time another ship, with permission granted by the government of Grand Turk, was also on the bank searching for treasure. During the winter of 1978, treasure hunters working with the permission of the Dominican Republic located an estimated $800 million worth of silver.

Before 1977 we spent many hours observing the humpbacks on this bank. Since Silver Bank is too shallow for large ships and is far from land, it was the least disturbed of the mating and calving areas in the West Indies. After making our way onto the bank with care, we could anchor the ship. Because most of a whale's life is conducted underwater, we joined the humpbacks there to find out what was happening. Diving with whales presents problems. Observation time is limited, and humpbacks can easily outswim a diver; therefore at best we could expect only brief glimpses of the underwater life of the whale.

To observe or photograph whales, it is necessary to get close enough to compensate for limited visibility but not so close as to influence their behavior in any way. Moreover, approaching close to a whale can be extremely dangerous. We have never known a whale to deliberately attack a diver, but even a soft tap from a fin or fluke could inflict serious injury. On one cruise, when a photographer was gently brushed on the hand by the barnacles on the tip of a humpback's pectoral fin as the whale turned, the resulting wound required stitches. Care must also be taken not to get too close to the flukes as the whale swims past; even a light blow could be disastrous. In the sea the whale, not the human, has the advantage.

We wait until whales are within 200 yards of the ship before we enter the water. The water is warm, but nonetheless is a relief from the hot sun. As we descend, the surface noise of wind and wave becomes a muffled roar. Rays of sunlight reveal millions of tiny suspended particles that create an underwater fog. Beneath us is a profusion of coral and sea anemones, with their venomous tentacles reaching out. Colorful sea fans sway in the currents; here and there are scattered patches of white sand.

We notice a vague white patch emerge from an area of darker water ahead. Motionless, we watch as the shadow takes a form; the white patch becomes the long pectoral fin of a humpback whale. Since the whale's underwater vision is about the same as ours, these white pectoral fins probably help the whale to keep in visual contact with other humpbacks. Two more animals appear and swim toward us. We begin to retreat and for a moment consider heading back to the ship, but curiosity triumphs over fear and we stay.

Underwater, humpbacks appear slim and agile as they move with their slow grace. Their flukes seem to flow, the long flippers undulate slowly, and their bodies are supple with great muscles rippling beneath glossy black skin. They look at us with eyes that seem incredibly small. We are awed by the physical power and presence of such huge creatures. As they turn to avoid us, we see how efficiently they use their long fins for turning and balancing. By rotating their fins, as much as 90 degrees, the whales are able to scull backward or forward, up or down. They can control each flipper independently and can even push themselves sideways using only one fin; with a simple movement of one or both flippers they can turn or bank sharply. Photographs are very deceptive because the fins appear to be stiff; actually they are flexible, almost as flexible as the human arm. As the humpback accelerates, the fins are swept back like the wings of a jet.

Because the humpback is the only whale with such long pectoral fins, scientists feel they must serve some special function. Why did the humpback whale alone evolve such distinctive flippers? We think there are three reasons. One is that long flippers allow the humpback to navigate in much shallower waters than other whales. Since humpbacks are relatively slow swimmers, the long fins are important too, for maneuverability. But perhaps the most important function is thermoregulation.

The humpback whale is the only baleen whale that regularly inhabits both tropical and frigid seas. This presents the problem of maintaining a constant body temperature whether the water is cold or warm. Thick blubber layers insulate the whale when it inhabits colder waters, but when it becomes overheated in warmer seas it cannot sweat to cool itself. The Russian cetologist, A. B. Tomilin discovered that the caudal, dorsal, and pectoral fins of whales are actually thermoregulatory organs. When the whale is too warm, the blood flow to the fins increases; the heated blood from the body travels to the capillaries in the fins. Because these capillaries are very close to the surface of the skin, the blood is cooled by the water. The cooled blood is then returned by the veins to the body. The longer the fins, the greater the number of capillaries; and the greater the number of capillaries, the more efficient the cooling process.

In cold waters, the whale's body temperature is maintained by restricting the flow of blood to the fins and by passing heat from the arteries to the veins before the blood reaches the skin's surface. Although the fins remain cold, the whale's internal body heat is maintained. We conclude that because of its long fins the humpback is able to regulate its body temperature in warm or cold waters more efficiently than any other whale.

Of 373 humpbacks observed by A. Wolman and C. Jurasz in Hawaiian waters during 1976, 75 percent occurred singly or in pairs: about half were in pairs, one-quarter solitary, and the rest in groups of three. Louis Herman's 1977 study of Hawaiian humpbacks yielded a higher percentage of groups of more than two. Among humpbacks observed in the Pacific feeding grounds, the Japanese scientist T. Nemoto found 50 percent solitary, 43 percent in pairs, and 7 percent in groups of three or more. It appears that humpbacks in the tropics and in feeding areas are constantly moving, joining, and intermingling with other individuals. At least in some cases the same pairings have been observed over several years, but this may reflect chance rejoining rather than bonding. Groups of three seem often to consist of two adults and a calf.

We can only speculate about the interactions among groups of animals, since the sex of the participants is usually unknown. On one occasion a group of three whales approached our ship at great speed and circled us. They repeated this maneuver several times, charging around the bow to the stern as if they were being chased. Two more whales appeared and joined the group, swimming crazily and making sharp turns around and under the ship. After a while they moved away, sending up huge blows. One soon separated from the other four, who went off together, stopping now and then to circle and come into physical contact with each other. Possibly this was a group of unattached males and a female engaging in sexual play. In Hawaii similar groups were visually sexed and found to be males and a female. Peter Tyack believes that these groups consist mostly of males chasing a female in heat.

After many days spent anchored on Silver Bank, we knew more about how a humpback spends its days in the tropics. The whales constantly moved around the bank, rarely staying in one place for more than ten or fifteen minutes except during a midday rest period from about 10 A.M. to 2 P.M. During this time they still moved, but at a much slower pace. In the morning they moved upwind, into the sun, along the reef. By afternoon they would be moving east. As they passed our ship they would stay close for a few minutes, then continue on their way around the bank. On a calm day, when they were swimming quietly, their blows could hardly be seen. When they were excited, and presumably breathing harder, the blows were much taller and we could sometimes make out the double blow. The groups we saw usually swam one behind the other, but Louis Herman observed five other formations in Hawaii: wedge-shaped, with animals spread symmetrically behind a leading whale; V-shaped, with multiple leading animals abreast; T-shaped; X-shaped; and diamond-shaped.

Although humpbacks feed among other species in the north, they do not interact with them. Recently in Hawaii, a humpback and a right whale were seen swimming together for several days, and a baby spinner dolphin was observed accompanying a humpback mother and calf for four weeks. But as far as we know, this sort of behavior is very unusual.

It may seem that the humpbacks spend their days only in swimming, but nothing could be further from the truth. The humpback is an acrobat and is noted for its acrobatic antics. Dubbed the "merry whale" by the Russians, it lives up to its reputation.

WM. C. KING

6. MERRY WHALE

One of the most distinctive characteristics of the humpback whale is the way it behaves. Other whales sometimes jump out of the water, stroke each other with their tail fins, follow ships, or slap at the water with their fins and flukes; only the humpback does all these things regularly.

Of all the great whales, humpbacks in particular seem to be drawn to ships. In the early days of navigation, an approaching whale made sailors uneasy. Ships were small and rather fragile, and contact with a whale could cause considerable damage—in some cases even sink the ship. Henry Hudson described his deliverance from a herd of whales during his last voyage, made in 1610: "Our course for the most part was betweene the west and the northwest, till we raysed the Desolations, which is a great iland in the west part of Groneland. On this coast we saw store of whales, and at one time three of them came close by us, so as wee could hardly shunne them; then two passing very neere, and the third going under our ship, wee received no harme by them, praysed be God."

Collisions with whales were common in the days of sailing ships. Vessels had no engine noise to warn off the whales, and certainly the whale population was infi-

nitely larger. One collision was recorded during the English explorer Martin Frobisher's third voyage to North America in 1578. A vessel of the fleet struck a whale "with such a blow, that the ship stood still and stirred neither forward or backward. The whale thereat made a great and ugly noise, and caste up his body and tayle, and so went underwater." In 1613 the French explorer Samuel de Champlain reported that his vessel passed over a sleeping whale and caused a large wound near the tail that bled profusely. These whales may well have been humpbacks, since they do rest motionless at the surface as if they were dozing. On several cruises we have had humpbacks snuggle up to our drifting vessel and seem to fall asleep, and Charles and Virginia Jurasz in Alaska have often seen humpbacks resting on the surface. The whales—low, dark, rocklike—move only occasionally, when they raise their heads to breathe. They move slowly, drifting with the tides and currents, and remain almost immobile for as much as two hours if left undisturbed.

Some felt the humpback harassed vessels deliberately. Sigurd Risting, in his book on the history of whaling, refers to an old book entitled *Kongespeilet* (The King's Mirror) in which humpbacks off the coast of northern Norway were said to be "naughty" toward ships: "One of their characteristics is to beat against the vessel with their swimming fins or to float and displace themselves in front of the vessel along its course; if the course is changed they swim in front of the boat again, and there is little choice other than to sail into them. But if this is done, the whale reacts by pounding the boat to pieces and pummeling all who were on board."

Others felt the whale demanded a sacrifice. On Richard Fischer's voyage to Cape Breton, the following incident, which probably involved a humpback, occurred off Newfoundland: "One thing very strange happened in this voyage: to witte, that a mightie great whale followed our shippe by the space of many dayes as we passed by Cape Razo [Cape Race, Newfoundland], which by no means wee could chase from our ship, until one of our men fell overboard and was drowned, after which time shee immediately forsooke us and never afterward appeared to us."

In 1658 the French explorer C. de Rochefort referred to his encounters with whales, probably humpbacks, in the West Indies. He wrote: "Ships are also sometimes accompanied for quite a long time by monsters which are of the length and breadth of a boat, and which seem to find pleasure in thus showing themselves." Even whalers were wary of the humpback, especially in herds.

Humpbacks occasionally follow ships today. In 1964 Carl Hubbs, a marine biologist, observed a humpback when he was trawling off Baja California; the whale accompanied the ship for about two hours, maintaining a speed of four knots. A fisherman from Digby Neck, Nova Scotia, while returning to shore after a morning of fishing, spotted a humpback heading straight for his boat. He swerved to avoid a collision, but the humpback kept coming and brushed up slowly alongside the boat. The startled fisherman tried to pull away, but all his maneuvers were in vain; the whale remained close alongside, leaving only when the boat had almost reached shore. The fisherman believed that the whale had mistaken his boat for a member of the opposite sex and set about courting it. Recently, for reasons open to the same sort of speculation, a humpback adopted a remotely operated submersible for several hours.

In his 1951 book *The Quest of the Schooner Argus*, Alan Villiers described several occasions when humpbacks encountered dory fishermen of the Azores on the Grand Banks of Newfoundland:

Then there were whales. I had not heard of these as a menace to dorymen, and when a family of humpbacks played around the ship for several days, I did not appreciate the wariness with which the dorymen regarded them.

While he was quietly hauling his long-line about ten o'clock in the morning, our Francisco de Sousa Damaso of dory 57, a green fisher from Ponta Delgada, suddenly saw the bulk of a whale rising beneath him in the clear sea. The whale's blubbery bulk was already tangled with his long-line, a circumstance which the whale had probably not noticed, though a surprised cod or two were flapping against his barnacled blubber. He failed to notice the dory, either. He came to blow almost right under it, toppled it until it all but capsized, and some of the gear fell out. Francisco Damaso, no stranger to whales, hurriedly cut his line and prayed. The dory slid off or the whale

Fig. 22. A humpback plays with a twenty-foot log, lifting it repeatedly with its head. *Photo by John Arsenault.*

slid away and, for a split second, an amazed doryman glared at an astonished whale. The whale blew, sounded lazily, and departed, still with the long-line trailing astern of it and Senor Damaso's morning take of cod going off to oblivion.

Villiers mentioned three other similar incidents in that same week, although the captains told him that such encounters with humpbacks were rare.

Whale-watching vessels off Cape Cod have been approached by humpbacks who lay alongside the ships, sometimes turning belly up. Although the whales remain close, they have not so far come into contact with any vessel. Observers on board the Dolphin III, a whale-watching boat operating off Cape Cod, have seen a humpback repeatedly surfacing under a 20-foot log, raising it up out of the water with its head (see Fig. 22).

Even though it is well documented that humpbacks will follow ships and occasionally make contact with them, why they do so remains a mystery. If, as some believe, the humpback possesses a superior intelligence, perhaps these acts are deliberately intended to harass the human enemy. Others might argue that the whale is actually a huge, blundering creature, unable to avoid even the largest objects, let alone to distinguish between a boat and another whale. It is possible, too, that the humpback is just displaying a natural curiosity. The fact is, we simply do not know; we can only speculate.

In all the years we have studied humpback whales in the North Atlantic and in the West Indies, at no time has one come in contact with any of our vessels. Whales often approached us, but we never knew why. It did not seem to matter whether we had the engines running or shut down, whether we were in a steel-hulled ship or a wooden-hulled boat; their behavior was the same.

Eventually, however, we discovered a pattern of activity. In the West Indies, we found that from about 10 A.M. to 2 P.M. we could not get close to a humpback, although we tried repeatedly. Even when the ship was anchored and completely silent, the whales kept their distance. They were quiet, and we saw almost no breaching or splashing. At other times of the day it was a different story.

In the early morning hours the banks would be alive with humpbacks breaching, finning, tail-slapping, and swimming about. Pairs and trios would approach the ship, swimming under and around it, often within inches of the hull. It made no difference whether we were steaming, drifting, or at anchor. By 10 A.M. all was quiet and the whales seemed to vanish. Occasionally we glimpsed a back or a fin, but that was all. At 2 P.M. the whales would reappear and approach the ship once more. They repeated the early morning pattern well into the night, for we heard splashes and soft puffing blows as we lay in our bunks. This made some of the crew uneasy, espe-cially those in the smaller 65-foot Eaglet II, a boat that a humpback could easily damage.

We found that others had observed this early morning activity. J. B. DuTetre, a mid-seventeenth-century French explorer, recorded probably the first description of humpbacks in the West Indies:

Whales are seen about these islands from the month of March to the end of May more frequently than in all the rest of the year. They are in heat and copulate at this time, and one sees them roaming about principally in the morning, all along the coast, two, three or four, all in a school, blowing and as if syringing from their nostrils two little rivers of water, which they blow into the air to the height of two pikes, and in this effort they made a kind of bellowing which may be heard for a good quarter of a league. When two males meet near one of the females they join battle and give themselves over to a dangerous combat, striking the sea so hard with their fins and tail that it seems as if they were two ships engaged with cannon.

These periods of activity may parallel the hours of intense feeding while the whales are in the north: that is, although the whales in the West Indies are not feeding, their daily behavior cycle may remain the same throughout the year. C. S. Baker and his colleagues characterize whales' behavior in feeding areas as noncompetitive and at times cooperative. In the tropics much male competition and thus agonistic behavior is involved in attempts to associate with females in estrus and with calves.

While the humpbacks swam around and under our ships, we often wondered what was taking place when they disappeared beneath the hull. On one cruise we found out. Our divers had spent several hours trying to approach a group of three humpbacks that seemed determined to remain just beyond their range of vision. At the magic hour of 2 P.M. the whales began to approach the ship in the familiar fashion, appearing and then disappearing, often quite close to the hull. Later one of the divers told us that as he was returning to the ship he had seen two adult humpbacks side by side and motionless just beneath the hull; it appeared to him as if they were looking up at the bottom of the ship. As soon as they saw the diver one whale took a position over the other, and in this formation they slowly moved away. Later in the same cruise another diver observed a similar incident involving a single animal; he also had the impression that the whale was looking up at the ship's hull.

Millais described humpbacks as "gay and sportive, frequently engaging in uncouth gambols" (see Fig. 23). Their chief antics are breaching, flippering, lobtailing, and spyhopping. Breaching, or jumping from the water, is the most impressive and startling. With no warning at all, a ponderous body explodes from the sea. Forty tons of bone and blubber defy gravity and hang suspended for an instant in a graceful arc, the long pectoral fins spread like wings. Then, the moment past, the whale crashes back into the water with an enormous splash. The sound can be heard for miles. Breaching is done throughout the year by both males and females, and young humpbacks have been observed breaching shortly after birth. No one knows why whales breach. At one time it was thought that breaching whales were trying to rid themselves of parasites, but this explanation now seems improbable. More likely the whale is responding to excitement or perhaps at times it is an aggressive display.

The first sighting of a humpback whale breaching is impressive, especially if one is close to the whale. There are many eyewitness stories. One of the more interesting, by Captain B. Hall, recounts an experience he had as a midshipman on board the H.M.S. *Leander* off Bermuda in the late eighteenth century. The crew was watching a humpback swimming around the vessel when someone suggested they take a closer look. A small boat was launched, and they rowed toward the whale:

All eyes were now upon us, and after a pause, it was agreed unanimously that we should run right on board of him and take our chance. So we rowed forward, but the whale slipped down, clean out of sight, leaving only a monstrous pool, in the vortex of which we continued whirling about for some time. As we were lying on our oars, and somewhat puzzled what to do next, we beheld one of the most extraordinary sights in the world; at least, I do not remember to have seen many things which have surprised me so much, or made a deeper impression on my memory. Our friend the whale, probably finding the water disagreeably shallow, or perhaps provoked at not being able to disentangle himself from the sharp coral reefs, or from some other reason of pleasure or pain, suddenly made a leap out of the water. So complete was this enormous leap, that for an instant we saw him fairly up in the air, in a horizontal

Hump-backed Whales at play.

Fig. 23. An impression of humpbacks at play by the English naturalist John Millais, first published in his 1906 book *The Mammals of Great Britain and Ireland.*

position, at a distance of at least twenty perpendicular feet over our heads! While in his progress upwards, there was in his spring some touch of the vivacity with which a trout or salmon shoots out of the water; but he fell back again on the sea, like a huge log thrown on its broadside, and with such a thundering crash, as made all hands stare in astonishment, and the boldest held his breath for a time. Total demolition, indeed, must have been the inevitable fate of our party, had the whale taken his leap one minute sooner, for he would have then fallen plump on the boat.

Captain Hall was so impressed with this sight that he wrote Captain William Scoresby, the famous English whaling captain, Doctor of Divinity, and Fellow of the Royal Society. Scoresby replied:

I have much pleasure in being able to speak to the point in attestation of the not infrequency of the exhibition of the high leaps which you witnessed, however ignorance might charge it as "very like a whale." Whilst engaged in the northern whale fishery, I witnessed many similar exploits of whales of forty or fifty feet in length, forgetting their usual gravity, and making these odd exhibitions of their whole form from head to tail. Certainly, I have several times seen whales leap so high out of water, as to be completely in air, which, reckoning from the surface of the back, could scarcely be less than twenty feet, and possibly might be more. I have at different times pursued these frolicksome fish, but they have always escaped.

We have watched thousands of breaches, but we have never seen an adult humpback clear the water while in a vertical position. Sometimes a breaching whale will ap-

pear horizontal to the water just before it lands on its side. We suspect that most descriptions of whales clearing the water actually refer to whales in the horizontal position.

There are several styles of breaching (see Fig. 24). Most humpbacks emerge vertically, then arch over and land on their sides. Others turn over and land on their backs, and occasionally some come up backwards, twist, and land on their bellies. The pectoral fins may be held closely to the sides, be spread like wings, or change position while the whale is in the air. Some humpbacks only breach once, but others leap twenty or more times in succession. Twice we saw a pair breach side by side.

It took many hours and rolls of film before we finally got an acceptable picture of a breaching humpback. Since there is no warning or indication where the whale will appear, fast reflexes are a must. Once the whale appears there is no time for camera adjustment. To further complicate matters, breaching usually occurs in the early morning or the late afternoon, when the sun is low and the light is poor. We probably have hundreds of pictures of splashes!

Tail-slapping or lobtailing is another antic of the humpback whale. The whale stands on its head in the water with only its tail flukes protruding into the air. It then slaps the water with the flukes, creating huge splashes and making a great deal of noise (see Fig. 25). Some whales stop after one or two slaps; others seem to go on forever. On one occasion we counted 40 tail slaps,

Fig. 24. Humpbacks have several styles of breaching. They may land on their sides or backs or fall forward; their fins may be spread or remain close to their body. For a moment their entire body may be out of the water.
Photos by Carol Price, Jeff Goodyear, and John Arsenault.

Fig. 25. This humpback is lashing the water with its flukes.

Fig. 26. A humpback lies on its side lazily waving its pectoral fin in the air. A tip of fluke can also be seen.

with slight pauses after every ten or twelve. Late one afternoon we counted 59 tail slaps made in succession by one whale. We have no idea how long this performance continued, since darkness fell and we could no longer see. As with breaching, there are several styles of tail-slapping. The humpback usually slaps with the ventral surface of the tail, but we have seen whales slapping with the dorsal surface. Sometimes the whole body stalk comes out of the water, and it may slap down at odd angles.

Some scientists think breaching and tail-slapping could be a means of communication. Both create a great deal of noise, which could easily be heard by other whales and thus perhaps be used as a warning in the way a beaver slaps the water with its tail to warn of danger. For the moment, at least, we can only speculate why humpbacks breach and lobtail.

Humpbacks also use their long flippers to stroke other whales, and to strike the water, themselves, and others. We watched one whale on its back at the surface throwing its flippers into the air one at a time, then crossing them over its body; next it would throw both flippers into the air at the same time. One group of four humpbacks flippered for three hours right beside the ship. Although we have never observed a whale actually striking or stroking another whale with its flippers, it has been reported that they will gently stroke or sharply strike each other during courtship. Louis Herman reports that swimming groups often touch each other with their fins; perhaps to maintain proximity. He observed one of a pair lying on its side with its pectoral fins extended (see Fig. 26). The other whale swam between the extended fins and was passively stroked from head to tail. Apparently physical contact between humpbacks can become quite intense. Hal Whitehead and Peter Tyack have noted Atlantic and Hawaiian humpbacks with worn, bloody dorsal fins and tubercles. Many more observations will have to be made before we can know exactly how they use their long fins to interact with each other.

Humpbacks have quieter amusements as well. One day, while we were anchored near a patch of sargasso weed on the edge of Silver Bank, three adult humpbacks joined us. They swam into the weed, lazily lifted their flippers, and let the seaweed dangle like tinsel. For almost twenty minutes they milled about in the weed, dipping and raising their flippers. One whale stood vertically in the water with only its head visible, the seaweed clinging wetly to its crusty snout. This position, with only the head out of the water, is called spyhopping. Whales may do this to "take a look around" or there may be some other totally unknown function.

Humpback whales breach, lobtail, and slap their fins while they are on their feeding grounds, but seem to be more active while they are in the tropics. This, after all, is their mating season.

7. MATING AND BIRTH

Much of what we know of the humpback's sexual life comes from studies of dead animals. The whale's sexual organs cannot normally be seen because they are concealed beneath a genital slit in both male and female, an advantage in streamlining. The female has two smaller slits on either side of the larger opening, which contains the nipples (see Fig. 27). We now know that by examining the ovaries of a dead female one can find how many times she has ovulated. This does not reveal the number of calves she has borne, but it does give scientists some idea of the reproductive cycle of females.

The ovaries of toothed whales resemble those of other mammals, but baleen whales' ovaries look more like those of birds and pigs. An adult female humpback's ovaries resemble a bunch of grapes. Each bump is a follicle, with a diameter of from 1¼ to 2 inches, and each follicle contains an ovum, a tiny, almost imperceptible speck. When a follicle matures and ruptures, the ovum enters the fallopian tubes and the whale is ready for mating. Usually one follicle ruptures at a time, but occasionally two or more may be discharged almost simultaneously, and this may result in multiple fetuses.

Fig. 27. The urogenital opening of a female humpback whale. The mammary slits can be seen just below the barnacles on each side of the opening. The so-called hemispherical lobe can be seen below the slit. It is only present in females and allows for quick identification of sex by divers. The anus is just below the lobe.

Once the egg is released, the follicle increases in size and forms new tissue, the corpus luteum. In most mammals this tissue is gray to yellow, but in rorquals for some reason (possibly due to diet) it is pink. If the egg is fertilized, the corpus luteum enlarges and produces a hormone called progestin, which stimulates the adhesion of the fertilized ovum to the wall of the uterus. Large amounts of progestin from the corpus luteum of whales were used by the pharmaceutical industry to help maintain human pregnancies. In cetaceans the corpus luteum continues to function throughout pregnancy but degenerates soon after the calf is born. If fertilization does not occur, the corpus luteum degenerates from ten to twenty days after ovulation; the pink tissue disappears, leaving a white connective tissue called a corpus albicans. In terrestrial mammals, seals, and sea lions the corpus albicans eventually disappears, but it never entirely disappears in whales. It is by counting the residual traces of corpora albicantia that researchers determine how many times a whale has ovulated in her lifetime.

The humpback whale usually ovulates once every two years, most often in the winter months but occasionally at other times of the year. Of the mature females studied in Western Australia 71 percent calved once every other year. The rest were divided into three groups; those that calved two years in succession, those that lost a calf and became pregnant again during the same season, and those that calved every third year. In multiple-year observations, Mayo has recorded calving at two- and three-year intervals by females off Cape Cod. One female calved two years in a row in Hawaii.

The sexual organs of the male also lie within the body. The testes, which are behind and lateral to the kidneys, are no different in structure from those of other mammals; even the sperm of the largest of the whales is no larger than man's. A small quantity of semen is present throughout the year, but there is a marked increase of sperm production during the winter months. Because of this and the fact that females sometimes ovulate at other times of the year, it would be possible for a humpback to mate successfully at times other than the winter months, but there is no evidence that this happens.

In form, structure, and position the cetacean penis is similar to that of bulls, goats, and stags. When retracted and flaccid, the organ assumes an S-shaped position inside the whale's body, but when the muscles slacken the penis becomes erect and protrudes. Since the sexual organs of male cetaceans are similar to those of even-toed ungulates, they probably copulate in the same way. Bulls and rams take no more than a few seconds to mate. If the encounter were that short between whales, it would be very difficult to observe mating.

Although the actual mating may be short, the foreplay is prolonged. E. J. Slijper, a Dutch cetologist, notes that "cetacean couples display great tenderness towards each other." Scammon, the whaling captain-naturalist from California, described the mating habits of the humpback in his book *Marine Mammals of the North-Western*

Coast of North America (1874): "In the mating season they are noted for their amorous antics. When lying by the side of each other, the megapteras frequently administer alternate blows with their long fins, which love-pats may, on a still day, be heard at a distance of miles. They also rub each other with these same huge and flexible arms, rolling occasionally from side to side, and indulging in other gambols which can easier be imagined than described."

Many whalers claimed to have witnessed humpbacks mating, but the first scientific publication describing their method was written by D. G. Lillie in 1910. He stated:

The Balaenoptera are said by whalers to copulate at the surface of the sea. The pair swim towards each other and turn slightly on the sides so that their ventral surfaces face one another. The male makes several passes at the female to insert the penis. When the pair first fuse together, the long axes of their bodies are parallel with the surface of the sea, but they curve up vertically at the end of the act. After copulation, the male is said to be exhausted and easily caught.

In 1948 two Japanese scientists, Masaharu Nishiwaki and K. Hayashi, described another method of copulation they had witnessed. The two humpbacks dived under the water, then surfaced together vertically with their ventral surfaces together, then fell back into the water. These whales were observed for three hours, and they repeated the behavior three times. The authors had previously heard from whalers that whales jumped above the water during copulation, but had doubted the story; there was no evidence, they add, that the male whale heaved a great sigh or became exhausted after the act. We doubt that the observed behavior was in fact mating, the more so since no mention was made of intromission.

During the mating season in the West Indies, we have seen pairs and groups rolling, splashing, and chasing each other. We suspect that some of this may have been courtship or mating activity, but we never could positively identify any act as copulation. We have also twice seen two humpbacks breaching simultaneously, but not in contact with each other. Louis Herman observed a group of Hawaiian humpbacks nuzzling a whale lying inverted in the water, but he could not determine the sex of any of the participants.

A fisherman from Nova Scotia told Carl Mrozek, a writer, that on more than one occasion he had observed humpbacks pressed closely together and rolling in the water between late August and October. The fisherman had also seen them breaching together during this same time, and stated that solo breaching usually began later in the season, after September 1. We really do not know whether humpbacks sometimes mate before or during the migration, only that most mating occurs in the tropics.

The bond between a mating pair is said to be much stronger during the mating season than at any other time of the year. There are many stories of one member of a pair standing by until the death of its companion, which behavior unfortunately usually resulted in its own death. Yet we are a bit suspicious about this "bond" being stronger during the mating season, since much evidence indicates that humpbacks are protective of their companions at all times of the year. On two separate occasions when Beamish was holding a humpback that had become entrapped in a cod net off Newfoundland, a second animal remained nearby for several days. Tomilin mentions several incidents involving the humpback's concern for its mate while in northern feeding areas. One incident, similar to the others, took place in the Bering Strait; a wounded male took the pursuing whaleship in tow, and an accompanying female would not leave him until he was killed by a second shot. Moreover, strong as the bond may be, Hawaiian researchers have determined that there is a great deal of interchange of male escorts with a female.

While some humpbacks are engaged in mating, others are awaiting the birth of their calves. The humpbacks' main purpose in migrating to the tropics is to provide their calves with a safe, warm environment. In the tropics the calves grow quickly, building their strength for the long migration back to the colder feeding areas. Calves born before or during migration have a lesser chance of survival.

Most of the little we know about pregnancy and birth in humpbacks comes from studies of dead whales killed before the ban on humpback whaling. Without these studies, we would not understand the animal's reproductive system, knowledge of which is vitally important to the future protection of the species.

Surrounded by two membranes filled with fluid, the humpback embryo is well protected within the uterus. The first membrane, the amnion, is similar to that which surrounds a human baby; the other, the allantois, lies outside the first as in humans. Because the gestation period is only eleven to twelve months, a short time for such a large animal, there is an exceptionally high rate of fetal growth. If fertilization takes place on February 1, by May 1 the young whale will be about one foot long with most of its organs formed. The head is still bulbous and arched downwards, the abdomen protrudes, but the form is that of an adult rorqual and all the fins are present. Still not formed are the grooves on the throat, the baleen, and skin pigment. The absence of pigment allows blood vessels beneath the surface to lend color to the epidermis, so that the fetus appears pink.

After six months of fetal development the first grooves appear between the pectoral fins and the umbilicus. Soon new grooves appear at the bottom of the mouth, where they eventually fuse with the others. Pigment first appears as a dark strip along the upper jaw, particularly at the tip of the snout, and as dark strips at the ends of the dorsal fin and flukes. Later irregular dark areas appear on the back and pectoral fins, but not until the fetus is thirteen or fourteen feet long does it develop the markings it will have at birth.

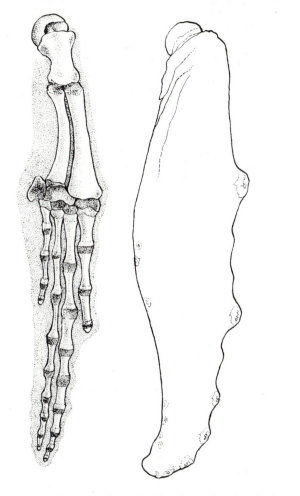

Fig. 28. The pectoral fin of a humpback showing the bone structure. *By William C. King.*

Early stages of fetal development reveal that humpbacks were once terrestrial mammals. An eight-inch embryo has rudimentary hind limbs, which disappear by the time it reaches eleven inches; fore limbs survive to a greater extent. The bones of the human arm and hand are in the flippers of the humpback (see Fig. 28). The bones of the flippers have four digits, or finger bones; humpbacks and other rorquals do not possess a thumb. During early fetal development a series of vestigial teeth form, evidence the humpback had a toothed ancestor. The teeth are small and bluntly conical, with 28 to a side in the upper jaw and 42 to a side in the lower jaw; they are reabsorbed and disappear before birth.

No one has witnessed the birth of a humpback calf, though a myth has it that the tail of the fetus emerges from four to six weeks before birth so that it can practice swimming. However, thanks to various studies of porpoises in captivity and to studies of dead whales, we do have some idea of what occurs. As pregnancy advances, the fetus is forced to assume a folded posture, molded by the position and shape of the uterus. Since the head, neck, and thorax of the fetus is the heaviest part of the body (see Fig. 29), the lighter, slimmer tail section is directed toward the genital opening and the calf is expelled tail first. The newborn calf surfaces for air as soon as the head emerges.

The birth of a humpback probably takes from 25 minutes to two hours. As the birth progresses, the fetal

Fig. 29. A near-term fetus. Throat grooves are fully developed, baleen can be seen in the upper jaw, and there is some pigmentation on the body. *Photo courtesy of Smithsonian Institution.*

membranes are shed near the cervix, with the result that the humpback is born without covering membranes. If the umbilical cord is stretched taut or becomes knotted, or if the blood flow is impeded in some other way, the fetus may be forced to try to breathe before birth, which will cause its death. The emerging calf pulls against the placenta and breaks the umbilical cord cleanly against the body wall. This leaves an empty tube of skin, about seven inches long, which previously enveloped the base of the cord (see Fig. 30). Occasionally the cord does not break and the placenta remains attached to the cord and the calf. The weight of the placenta may prevent the calf from reaching the surface for the air necessary for its survival.

Once the calf is born, it cannot breathe underwater because the stimulus for opening the blowhole is air.

Without air in its lungs, the calf tends to sink, so the mother may push it to the surface for its first breath.

Soon after the calf's birth, the mother expels the placenta. The cetacean placenta, the tissue that attaches the embryo to the wall of the uterus, does not become fused with the mother's tissue. The respective vascular systems are separated by two capillary walls and two epithelial layers. This is why whales do not lose as much blood as humans when they give birth.

The newborn humpback must surface frequently for air, keep warm, and follow its mother. Humpbacks are born with their eyes open, with good hearing, and with the ability to swim. Although the newborn humpback's dorsal fin and flukes are flaccid and do not become rigid for some time after birth, the calf swims easily, if awkwardly, from the moment of birth.

Fig. 30. Mother and calf. The attachment near the umbilicus of the calf may be a tube of skin left when the umbilical cord separated; if so, this calf was born a short time before the photograph was taken. *Photo by Dave Woodward.*

Nursing begins almost immediately. As we have seen, cows have two slit-like openings, on either side of the genital opening, in which the nipples are recessed. Because it must surface for air the calf cannot nurse continuously. When the calf wishes to feed, the mother's nipples protrude and the calf grasps the teat between its tongue and upper jaw. The mother squirts large quantities of milk down the calf's throat. Each nursing session lasts only a few seconds. Calves will nurse about forty times throughout the day and night and consume about three gallons of milk at each feeding. The daily intake of 120 gallons may seem enormous, but remember that newborn humpbacks are about fifteen feet long and weigh 1,500 pounds.

Humpback milk is creamy white, sometimes with a pink tint. Its odor can be slightly fishy, and it tastes like a mixture of fish, liver, milk of magnesia, and oil. Because it is so concentrated it looks like condensed milk, and indeed its water content of only 40–50 percent is much less than the 80–90 percent water content of domestic animals' milk. Its fat and protein contents are 20 percent and 12 percent, respectively, but as weaning approaches, the fat content increases to 29–40 percent and the protein content to 11–13 percent. The milk of the humpback whale has almost no sugar.

The sex ratio at birth favors males slightly. Twinning is infrequent. Among 1,449 pregnant whales taken in the Antarctic in 1949–1950 and 1955–1956 seasons only four cases of twinning were recorded, and Australian studies carried out since 1949 have recorded only one case of twins. The likelihood of both twins surviving is extremely small. As far as we know, no one has ever observed a humpback with twin calves.

Although we have never seen a humpback calf nursing, we came very close on one occasion while snorkling on Silver Bank. We were just beneath the surface as a mother and calf approached. The calf dived and began to poke and prod its mother's chest just beneath her pectoral fin. Persistently it explored her rubbery skin with its snout, gradually moving toward her tail. We hung motionless in the water, hardly daring to breathe. Just as the calf reached the area of the mammary slit, the mother jerked away. The calf quickly took its position above her right flipper, and the pair moved away.

A calf remains dependent on its mother for about a year (see Fig. 31). Weaning begins at five or six months, although yearling calves have been observed trying to nurse. Weaning is probably a gradual process. There is some evidence that calves begin feeding during their first summer in the north, but it is likely that they obtain most of their food from their mothers during their first year. By the time a calf makes its second migration it is completely on its own, though it may swim in a family group for another year or two. Young humpbacks grow quickly, doubling their length by the end of the nursing period. Their growth then continues at a slower rate until they reach physical maturity at about ten years of age. The lifespan of a humpback whale is estimated to be 30

Fig. 31. During the first year on the feeding grounds the humpback calf learns to feed but depends mostly on its mother's milk. This calf surfaces with its mother as she feeds in the North Atlantic.

years or longer. Sexual maturity is reached at four to five years, and females who live out their lifespan give birth to about twelve calves.

Many have written about the behavior of female humpbacks and their calves. The bond between mother and calf is strong; a female will risk her own life rather than abandon her calf. The calf, in turn, will not abandon its mother as long as she lives. Whalers often took advantage of this situation. Charles Nordhoff, in his book *Whaling and Fishing* (1856), wrote of the humpbacks in Antongil Bay, Madagascar:

The females of these whales, as well as the right whale, frequent bays and shallow waters yearly, when their time of calving comes on, to drop their young, remaining in the smooth waters until the young leviathan has gained strength sufficient to shift for himself on the broad ocean. These occasions are taken advantage of by whalemen, and great numbers of the fish are slain annually in the many unfrequented bays of Africa and South America.

Most whalemen were simply doing their job, although at times they did not like what they were doing, as Nordhoff makes clear:

The mother whale seemed solicitous only about her calf. She would fondle it with her huge snout, and push it along before her. She would get between it and the boats, to keep it out of harm's way. She would take it down with her, knowing that on the bottom was the safest place. But here the little one could not obey her. It was forced to come up to breathe at least once every two minutes, and by this means, even had we not been able to tell by the strain of our lines, we knew at all times where the old whale [was].

Never did mother, of whatever species, display a more absorbing affection for her young than did this whale, and there was scarcely one in the pursuit, but felt as though we were taking a dishonorable advantage of her.

American and European whalers were not the only ones affected by the plight of mothers and calves on the whaling grounds. On a small island off the coast of Japan there is a Buddhist temple; on the temple grounds is a shrine and near it a monument over seven feet high, marking the spot where whale embryos taken from the mother during flensing were wrapped in straw mats and buried.

For centuries a net fishery for humpbacks and other whale species was operated off the coast of Japan. The whalers, influenced by the strong affection shown by the female whales for their calves, asked that the shrine be built. When the embryos were buried, the net whaling groups made valuable offerings and a requiem was said for the repose of the mother whales' souls. The sutras chanted by the monks were the same as for humans. Each whale was given a posthumous Buddhist name, which was registered with the date of capture. The requiem is still said today.

Mother humpbacks fight fiercely for their calves, often

grasping them with their flippers and taking them down with them. Even when a mother is mortally wounded her only concern is for her calf, and she uses every means she has to keep it from harm.

Calves are sometimes just as protective toward their mothers. Millais tells us that in June 1903 Captain Nilsen of the whaler *St. Lawrence* was hunting in Hermitage Bay, Newfoundland, when he came up to a huge cow and her calf. After harpooning the mother and seeing that she was exhausted, Captain Nilsen ordered a boat lowered for the purpose of lancing. When the boat approached the wounded whale, the young whale kept moving around the body of its mother. Every time the mate tried to lance the whale the calf intervened, and by smashing the water with its tail whenever the boat approached, it kept the whalers away for half an hour. Finally the boat had to be recalled for fear of an accident, and a fresh bomb harpoon was fired into the mother, causing her instant death. "The faithful calf now came and lay alongside the body of its dead mother, where it was badly lanced but not killed. Owing to its position it was found impossible to kill it, so another bomb harpoon was fired into it. Even this did not complete the tragedy, and it required another lance stroke to finish the gallant little whale."

"Battle of the Summer Islands," a poem by the seventeenth-century English poet Edmund Waller, describes an incident that occurred in Bermuda during the hump-back season. A whaling ship encounters a mother and her calf. When the calf eludes the crew's lances, the men attack the mother and wound her, whereupon the calf "hastes to her aid":

> The men amaz'd, blush'd to observe the seed
> Of the monsters human piety exceed!
> Their courage droops, and hopeless now they wish
> For composition with th' unconquered fish;
> Not daring to approach their wounded foe,
> Whom her courageous son protected so.

The story ends with the whales escaping on the rising tide.

Some whalers did not take calves, not realizing that it would have been much kinder to kill the calf along with the mother. Nordhoff understood this when he wrote, "I thought it would have been the part of mercy to kill the calves, since they would most likely starve to death." He had once seen a motherless calf, half-starved and desperately trying to find a source of food, trying to nurse from a little bull humpback. The bull would have none of this; as the calf approached, he "would wheel round and strike at it with his flukes, sometimes hitting, but oftener missing it. His short loud spouts showed clearly that he was in some consternation, and did not quite understand the maneuvers of his troubler."

Today, despite the end of whaling in most parts of the world, a calf is still helpless if its mother dies. In 1975, a

MATING AND BIRTH

young, thin humpback calf was seen swimming alone for about three days near Molokini Island, Hawaii. The calf approached a scuba diver taking photographs and allowed him to stroke its pectoral fins and underside. Several days later, the calf was attacked and killed by sharks. Although there were many other whales in the vicinity, the calf made no attempt to join them for protection, nor did they seem concerned about the calf's plight.

One pleasant aspect of a cruise is watching young whales. We often encountered humpback mothers and calves weaving around the coral heads in shallow water. Several times, as we watched a mother and calf underwater, we were startled to see a third whale lurking below us, just beneath and behind the pair. It would remain on guard until the mother and calf began to move off; then it would rise and take the lead.

Louis Herman noticed this same behavior among Hawaiian humpbacks. He reported that the third whale "usually remained at lower depths, most often to the rear of the pair," but occasionally, in the presence of a disturbance, swam in to shelter the calf between itself and the mother. Two Hawaii researchers, Deborah Glockner-Ferrari and her husband Mark, have visually sexed some of these escorts as male.

Diving among whales without disturbing them is difficult. We soon learned that the best way to observe mothers and calves is to stay in one place and let them

come to us. Calves were allowed to come quite close to a diver or a ship as long as no movement was made toward them. One female with a very tiny calf approached the ship and allowed the calf to come even closer on its own. Females with calves often came to the ship and swam around and under the hull, the calf hovering just above the mother's body. Occasionally we noticed an adult disciplining an overzealous youngster by pushing it away from us. Humpback calves, like all young animals, are playful and curious; humpback mothers, like all mothers, are sometimes stern and impatient.

The calf, like its mother, has its long pectoral fins swept backward. Calves work very hard to match the mother's pace, using two strokes of their flukes to the mother's one. Each calf seems to have a favorite swimming position. Some prefer the area just above the mother's right flipper, others the left: "right-handed calves," as we referred to them, were more common than "left-handed" ones. As far as we could see, they did not switch sides. One calf remained on its mother's right side for over two hours and maintained this position as the pair swam away (see Fig. 32). After observing a group of humpbacks Alan Villiers noted: "We could clearly see the baby clasped in its mother's flipper, generally on the lethand side, as the family came to the surface." Other observers have reported, however, that a mother will routinely place herself between a calf and a ship.

Smaller calves often swim just above the mother, espe-

Fig. 32. Calves seem to have a right- or left-handed preference just as humans do. This calf preferred a position over the right fin of its mother and was never observed swimming on her left side. *Photo by Dave Woodward.*

cially when they are moving rapidly. Sometimes calves actually hitch a ride by resting on the mother's back, and are occasionally lifted clear of the water (see Fig. 33). There seems to be close communication between mother and calf. The young whale duplicates each turn the mother makes exactly, as if tied to her with an invisible string.

The smaller calves we saw seemed very clumsy. Lunging up out of the water nose first, they would blow, then fall back on their throats with a little splash. They had not yet learned the smooth, rolling action of the adults. Calves rarely stayed below for more than a few minutes, and when they did surface they blew two or three times to the mother's one. Many calves had white scratches on their backs and sides, probably because of rubbing against the barnacles on the mother.

The young humpback very soon attempts to imitate the behavior of the adults. We saw calves breaching, flippering, spyhopping, and tail-slapping. One adult female created a great turbulence by lashing the water with her tail, each smack making a sharp crack in the stillness. As we watched, another tiny tail appeared. The calf fiercely began its own imitation of the mother's behavior, the little tail flopping into the water with a soft plop. Some older calves nearly cleared the water when they were breaching. The smaller ones could not leap as far and emerged only halfway before falling back on their throats.

Mothers seemed to enjoy this play as much as calves. Often a mother would rest at the surface while her calf slid and rolled across her back. Sometimes she would lift the calf out of the water by rising as it swam over her, and it would slide down her side and splash into the water; sometimes she would use her fins to push or guide the calf. There are stories of humpback mothers grasping calves in their flippers and tossing them in the air, but we have never seen this done.

We have made many sound recordings in the presence of mothers and calves, but have never heard them producing sounds. This does not mean that they never use sound communication. The humpback whale is an extremely vocal species, especially during the mating season when adults produce a complex and lovely call. Probably the first sound a calf hears is this hauntingly beautiful song.

Fig. 33. Calves often rest briefly on their mothers' backs as this one is doing. *Photo by Dave Woodward.*

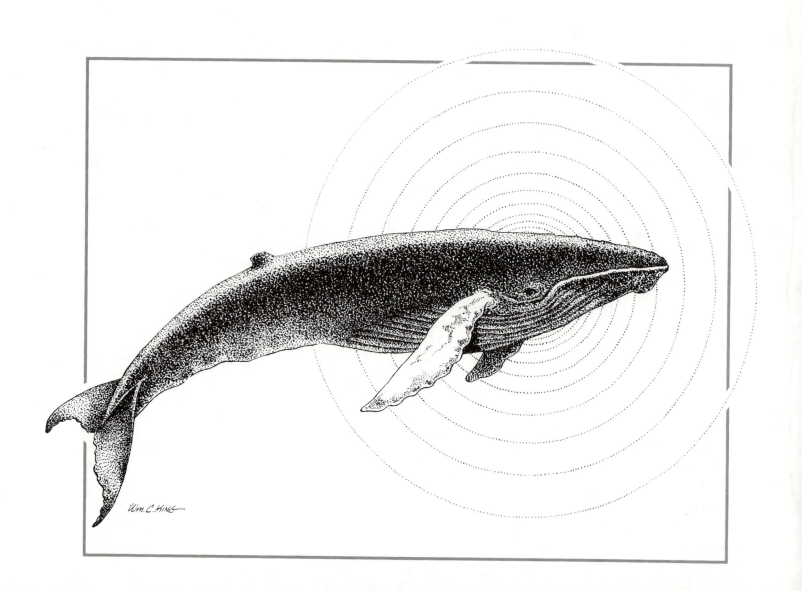

Wm. C. King

8. SINGING WHALE

One of the most important parts of our work with whales is recording their sounds. The humpback whale produces a variety of sounds, some of which probably help the whale keep in contact with others and may aid in navigation. What is less clear is whether baleen whales echolocate their food in the same way porpoises and toothed whales do. To echolocate is to emit sounds that bounce off nearby objects and to determine the location of an object according to the echo it produces.

Scientists demonstrated the echolocation ability of porpoises by placing an animal in a tank, blindfolding it, and then waiting to see if it was capable of finding its food or other targets. Recordings made during these tests showed that the porpoise emitted a series of high-frequency pulses that produced echos to which the porpoise responded with amazing speed and accuracy.

Since large whales could not be placed in a tank, experiments had to be carried out at sea. Sound recordings of sperm whales, the largest of the toothed whales, showed that they emitted high-frequency pulses similar to those made by porpoises and echolocated food in much the same way. Baleen whales, by contrast, only occasionally produced repetitive pulses that lasted a few seconds, and these pulses were much lower in frequency than those of the toothed whales. The results were promising but not conclusive.

In July 1976 a unique opportunity arose to study a baleen whale in captivity. An adult humpback whale in good condition became entrapped in a fisherman's net a few hundred yards off the coast of Newfoundland. Usually these unfortunate whales drown or are destroyed by fishermen, but in this case Peter Beamish, a scientist working at the Bedford Institute, Nova Scotia, was notified. After freeing the whale from the net, he tethered it with a padded harness. A holding pen was constructed with nets, and the whale was allowed to swim about in the enclosed area. A maze of aluminum poles was arranged within the enclosure (see Fig. 34), and during the next several days a series of tests was carried out to determine whether the whale could use sound to avoid the barriers.

The first tests were conducted in daylight, and the whale had no trouble avoiding the barriers. Since the whale remained silent during this time, Beamish assumed that it was navigating visually. For the next tests the whale's eyes were covered with a foam-lined blindfold. Still no sounds were heard, but this time the humpback blundered into the barriers. Further tests were

Fig. 34. A humpback rescued from a fisherman's net off Newfoundland attempts to swim through a maze while blindfolded. The white pectoral fins are clearly visible in the light fog.

made: at no time did the whale use sound while navigating the maze, though it did produce a few sounds at other times. Eventually the humpback was released unharmed.

These results seem to imply that humpbacks must rely on vision and do not use high-frequency sound for echolocation in the same way porpoises do. As we have seen, some scientists believe that baleen whales use lower-frequency sounds to locate larger objects, such as a ship, or to identify bottom topography, but further research will be necessary to verify this hypothesis.

Another sound humpbacks make is a strikingly beautiful song. Though few people have heard about it and even fewer have heard it, whale music can be plaintive and haunting, if at times demonically noisy.

For thirteen years we have recorded and studied humpback songs from the West Indies, Tonga, Bermuda, the Pacific coast of Mexico, Hawaii, and the Cape Verde Islands. Humpbacks begin to sing just before or during migration to the breeding grounds in the tropics, and they continue to sing throughout the breeding season. Recent recordings made in the North Atlantic and in Alaska suggest that they begin to sing the full song just before migrating to the tropics, and that only portions of the song are sung throughout the summer months on the feeding grounds. We have learned a great deal, but many questions remain. Humpbacks have been singing for thousands of years, but humans are just beginning to listen.

No one knows why or when the humpback developed its complex call; we do not even know when humans first heard these voices from the sea. Aristotle wrote that porpoises were capable of making sounds when at the surface, but he thought the larger whales were mute. We know humpback songs can be heard through the wooden hull of a boat. Since the ancient vessels that carried men at sea were frail craft, constructed of wood or hides, it is likely that whale sounds were heard, but how were they explained? The answers may lie buried in myth and legend.

Strange sea sounds are often associated with enchanted islands, and islands occur frequently in Irish legends. In the eighth century Maeldune, on a voyage in search of his father's murderers, came upon enchanted islands. "And as they went on they heard in the north-east a great shout and what was like the singing of psalms. And that night and the next day until nones, they were rowing till they could know what was that shout or that singing. They saw an island having high mountains full of birds."

The Irish folk hero Usheen heard the singing of birds on the Island of Youth. In another tale Bran the Blessed, after being wounded with a poisoned dart, ordered his men to cut off his head, assuring them that the head would be a pleasant companion since it would be able to talk! After removing Bran's head, the men set off for an unknown island, Harlech. At some point in their voyage they heard birds singing: "The songs seemed to them to be a great distance from them over the sea."

Perhaps the most famous Irish legend is that of Saint

Brendan (or Brandon), who supposedly journeyed for seven years in search of Paradise. On Easter, one year after leaving Ireland, Brendan and his companions reached an island filled with birds whose "happy singing was like the noise of heaven." At another point in their journey they approached another island around whose coast moved three choirs singing hymns.

The tales sound fanciful, but there is little doubt that some of these voyages actually took place. Some authorities believe the legend of Brendan's voyage is a composite of many voyages made by Irish monks of the period. Tim Severin, an expert on exploration, recently completed a trip from Ireland to Nova Scotia in a hide-covered boat (curragh), showing that Brendan could have reached the coast of North America. The common thread in all of these tales is the association of singing birds with enchanted islands. It is possible that some islands did have large populations of birds. However, to men of this time birds were the only animal known to sing and therefore birds offered the only plausible explanation for the songs they heard.

The *Odyssey* tells of what may be the most famous enchanted island, the island of Circe. But in Homer the source of the singing is not birds but sirens, who are half woman, half bird. In another siren story, the swan children of Lir, turned into swans by their stepmother, spent 800 years on a rocky island in the Atlantic, from where sailors and fishermen often heard the sweet notes of their songs.

Mermaids (half woman, half fish) are also used to explain sounds of unknown origin. In yet another Irish legend, that of Liban, a girl asks God to give her the form of a mermaid. The wish is granted, and Liban roams in the sea for 300 years. At some later date Beoan, son of Innle, was sent by Comgall to Rome to talk to Pope Gregory and bring back rules and orders for the Irish Church. "When he and his people were going over the sea they heard what was like the singing of angels under the curragh." The singing was of course attributed to Liban.

We can never be certain if any of these sounds were made by the humpback whale. But we do know that some portions of the humpback song sound amazingly like birds chirping, and no one who has ever heard a group of humpbacks chorusing would find any difficulty in likening them to a heavenly choir.

At times singing humpbacks do not sound heavenly. The low tones of their songs can be especially eerie and haunting, but here again folklore offers interesting grounds for speculation. Horace Beck, in his book *Folklore and the Sea*, mentions "one island in particular that qualifies as major or demon-haunted." Bermuda lies hundreds of miles from the mainland, rises precipitously from great depths, is frequently shrouded in rain, and has harbors that are difficult to enter. Six months out of the year humpbacks sing near its shores. In 1599, the English geographer R. Hakluyt described Bermuda as "a most prodigious and enchanted place."

We must keep in mind that less than 40 years ago,

when whale sounds were first recorded, speculations about their origin ran the gamut from enemy submarines to aliens from outer space. More recent evidence reveals that humpback sounds heard through the hull of a boat can easily be attributed to some other source.

Nordhoff, while whaling during the mid-nineteenth century, told of how a humpback would sometimes get under the boat and "utter the most doleful groans, interspersed with a gurgling sound such as a drowning man may be supposed to make." One morning a steward ignorant of such sounds went into the forecastle and was startled by a most unearthly groan that seemed to be coming from Nordhoff's bunk. He hurried off and informed the cook that the ship was haunted. When the cook ignored him, he went to the other crew members, who accompanied him to the forecastle. They all heard the groans, became terror-stricken, rushed to the deck, and made off in the dory to a neighboring ship.

In vain Nordhoff tried to explain that the sounds had come from a humpback whale; they were convinced that since the groans came from Nordhoff's bunk, an accident would soon befall him. "So eagerly does ignorant humanity swallow the most erroneous humbug, if there is only something supernatural about it, that of the sixteen men who had probably heard the same groans dozens of times not one could now be convinced, by reason or ridicule, that those in question owed their existence to a natural cause."

Perhaps this incident illustrates how, if the source is not visible, a natural sound can become the basis for superstition, myth, and legend. It is curious that although many whalemen must have known that the humpback produced sounds, few ever mentioned the fact in their logs, books, or journals. A unique entry in the 1883 log of the bark *Gay Head* states that a "singer" was caught on August 15 in Parita Bay (on the west coast of Central America). Since the date and location correspond with the movements of the humpback, this appears to be the earliest document on record that refers to humpback sounds as a song.

Not until World War II did people actually begin to listen systematically to underwater sounds. During this time SOFAR (Sound Fixing and Ranging) stations were set up by the U.S. Navy to listen for approaching enemy submarines, and submarines were also equipped with listening gear. Soon reports began to come in of strange unidentified sounds. Zoologists were consulted, but they knew very little about sounds in the sea. For six years the seas were monitored and some sounds identified, but very little of this knowledge reached the public for security reasons.

Submarines were so noisy at this time that most had to be equipped with filters that eliminated lower-frequency sounds; these filters, of course, cut out many whale sounds. After the war a hydrophone, the BQR, was developed and was so sensitive that a much quieter submarine had to be designed. Filters became unnecessary, and most whale sounds could be heard. Many sounds still remained unidentified, but suitable equipment for studying sounds in the sea was at last available for scientists.

Sounds in the low audiofrequency region produced by marine life have been recorded at the U.S. Navy SOFAR (Sound Fixing and Ranging) Station, Kaneohe Bay (Hawaii). These sounds have levels of the order of one microbar and many have a rather musical quality. There is a marked seasonal variation in the production of these sounds, with the early spring months being the period of most frequent occurrence. This variation is coincident with the seasonal variation of whales in the area, and this feature, plus the characteristics of the sounds themselves, has led to the belief that they are produced by whales.

This brief paragraph by O. W. Schreiber was published in the *Journal of the Acoustical Society of America* in 1952; it marks the beginning of our awareness of the song of the humpback whale. Not until 1969, however, did the whale biologist Roger Payne report that sounds made by humpbacks around Bermuda were long, patterned sequences, and that they were repeated over and over. No other whale is known to produce such a complex song.

While Payne was working in Bermuda, we were recording humpbacks around Puerto Rico and the Virgin Islands. We found that they also produced a long, patterned call. Our ship, the University of Rhode Island's research vessel R/V *Trident*, was equipped with a directional hydrophone attached to the end of a shaft that could be lowered from beneath the ship's hull. By turning the hydrophone, we could locate the direction of the strongest signal, move the ship in that direction, stop and listen again, take a bearing, and eventually come upon the calling whale. This method allowed us to locate calling whales quickly and accurately and to follow them throughout the night.

From these early recordings we discovered that single calling whales were always separated from groups of other humpbacks. We used a second sound track on our tape recorder to describe the whale's behavior as we were recording the sounds. Eventually we were able to correlate a particular sound with surfacing. This allowed us to anticipate the whale's surfacing; it also gave us a clear beginning and end to the call, which lasted anywhere from six to twenty minutes and was repeated again and again. One humpback we recorded repeated its call continuously for 24 hours and was still calling when we left it. In over a hundred encounters with single whales, we found only two that were not calling.

We hypothesized that the singers were male, but in order to find out we had to develop some method of sexing the whales at sea. Sexing whales by viewing their undersides was possible but difficult, time-consuming, and dangerous. John Crenshaw, a geneticist friend, suggested that we use the same technique used on humans to determine sex, the examination of cells obtained from mouth scrapings. This seemed an excellent idea, but how were we to obtain mouth scrapings from whales at sea? Nor were we sure that this technique would work with whales.

To test the theory we visited the beluga whales in the New York Aquarium. Trained to take food from the hand, they obligingly submitted to having their cheeks scraped; we also took scrapings from their skin. Very early in the embryonic development of the female, one of the X chromosomes becomes condensed into a dark stained mass called the chromatin body; the presence or absence of this mass would indicate the sex of the whale. The samples obtained from the belugas showed us that the technique worked, at least on toothed whales.

Our next problem was to find a way to obtain cell samples from calling whales at sea. One day Roy Mackal visited our lab. For an investigation of the Loch Ness Monster (he sought to determine if it existed), he had designed a dart-gun device to obtain skin samples of the creature. We developed an adaptation of it for our purpose, a hollow biopsy dart about six inches long fired from a modified harpoon gun. When the dart hits the whale's hide, a 1/4-inch piece of skin is retained inside its core. To guard against infection the tip of the dart is always covered with antibiotic salve. A piece of line is attached to the dart so that it may be retrieved (see Fig. 35).

Although the method seemed simple and efficient, there were difficulties. The lone singing whales were not easy to approach, as whalers knew; funds limited our time at sea; and weather conditions were often impossible. We managed to obtain samples from only two callers, both male. Most of the single animals we encountered in the Caribbean were producing the characteristic humpback call or song. It is now known that it is the males who sing.

Three basic kinds of sounds are produced by the humpback: high, low, and changing or modulated frequencies. These three sounds, alternated and varied, make up the song. Often the sounds can be grouped into six themes, but in some years we have identified only four or five themes. Listening to the call, especially when several whales are singing simultaneously, one gets the impression of a complex variety of sounds.

In fact, there are about seven basic types, with one to several variations of each. When we began analyzing these songs, we gave each sound a descriptive name so that we could discuss their differences: moans, cries, chirps, yups, oos (and other changing-frequency sounds), surface ratchets, and snores. Of the seven sounds only chirps and surface ratchets are missing from some calls, at least to date; the others are always present in some form, though they are put together in different ways. We have recordings from twelve consecutive years, two oceans, and five geographic locations, and so far there have been no exceptions.

The low-frequency sounds can be grouped into four types: moans, groans (a type of moan), snores, and surface ratchets. The modulated or changing-frequency sounds are the largest group and include wavers, oos, ees, whos, wos, and foos; other sounds that change frequency more abruptly are the yups, and the similar mups and ups. The high-frequency sounds are cries and chirps.

Moans or groans can be long, short, wavery, pulsed, or in two parts. There are several varieties of cries; chirps exhibit less variation. Variations of the snore include

Fig. 35. Howard is attempting to take a skin sample from a humpback by firing a dart from a modified harpoon gun.

short snores, long snores, elephant snores, lion snores, and two-part snores. The changing-frequency sounds have the greatest number of variations; they can be short, long, pulsed, or in two parts. The surface ratchet, which does not always occur, has only one or two variations. In the West Indies we found that the surface ratchet was associated with surfacing, but this sound was absent in the Pacific humpback's call. Even when the surface ratchet is absent, each call has a particular sound or sounds associated with surfacing: that is, surfacing invariably occurs at the same point in the song except when a disturbance causes the whale to surface earlier.

The term "song" is perhaps misleading, suggesting as it does an utterance intended more for aesthetic value than for communication. A humpback's song, like a bird's, is a long, patterned sequence. When natives of Tonga were asked if they were aware that the humpback sang, they answered very matter-of-factly that the humpback does not sing, it "talks." This is perhaps closer to the truth.

It is reasonable to suppose that whales, like other animals, have evolved methods of communicating that are particularly responsive to the unique environment of the species. Whales' environment is the sea. Vision is limited underwater, but sound carries well and at certain depths in the ocean may carry for hundreds of miles. Since the humpback does not have vocal cords, the source of its sounds is not known. The best guess is that they are produced by the various valves, sacs, and muscles found in the whale's larynx and in its respiratory tract.

The whale's ear, a tiny quarter-inch slit that is barely discernible and often wax-filled, is more efficient than it may appear. Humpbacks, like all whales, have standard mammalian ears modified to accommodate their aquatic life. The middle and inner ear are set in a very dense bone, the tympanic bulla. Since water is denser than air, the middle ear of the whale is more rigid and has greater mass, which allows optimal transmission of waterborne sounds to the brain. As a result of evolution the inner structure of the mysticete ear is particularly sensitive to low-frequency sounds, although its sensitivity to high-frequency sounds is probably also good. Dolphin whistles range from 2 to 15 kilohertz (Khz); humpback sounds reach only about 4 Khz, with the majority of sounds below 1 Khz. The human ear is most sensitive at 3 Khz.

Along with its keen sense of hearing, the whale has the capability of analyzing and interpreting the sounds it hears. Because whales have large brains there has been speculation that they are as intelligent as human beings or more so. In particular, it has been suggested that their song may imply communication of a high order such as exists in human language. This is surely not so. Although the song probably contains considerable information, everything suggests that it is information of a rather simple kind.

Whatever the meaning of the humpback song, the whale constantly repeats the same message over and over. Not only are individual sounds repeated, but groups of sounds are repeated, as is the entire song. Birds sing in much the same way, varying the order and linkage of a similar number of syllables into various song types. But the songs of some birds have many more syllables than the humpback whale's song, and their pattern is not so fixed.

Katy Payne, a biologist, found that Bermuda humpback songs change to some extent each year; this year's song is always slightly different from last year's. All the changes take place during the singing season, not between seasons, and all the whales change their song together as the season progresses. The changes are not drastic and may involve only a slight change in two or three of the sounds. The basic frequency patterns remain the same; for example, moans may change from long to short or wavery, but they are still moans. The reason for these changes is not understood. It may be that a new song each year is more stimulating to the females, or that one dominant male decides the song for the year; but for whatever reason, most humpbacks in a breeding area sing the song for the year, including the seasonal changes.

Since the humpback song was discovered to occur in the tropics during the mating season, little attention was paid to the sounds being produced on the feeding grounds. Recently, Nancy Reichley has recorded and analyzed a variety of sounds produced by humpbacks in the Cape Cod area during the summer months. She found that humpbacks have quite a repertoire of sounds consisting of about 21 basic sound types with at least 84 variations. The sounds were similar in type to those recorded in the tropics, and included moans, grunts, yups, cries, and chirps. Very few of the sounds were arranged in continuous patterned sequences similar to the song of that year in the West Indies, and most were either produced sporadically or continuously with no songlike pattern. Similar types of sounds have been recorded in Alaskan waters. Although the functions of humpback sounds from feeding areas are poorly known, they are undoubtedly used in maintaining contact when the animals are out of sight, in socialization activities when close, and may somehow be used in feeding.

Many animals, including humans, use pitch or frequency to communicate information, and this may be an important element in the whale's message. Since all humpback songs include a combination of frequencies, the humpback could be using different frequencies to send different messages to different individuals. Thus if all singing whales are in fact males, the low frequencies could be a threat to other males, the high frequencies an inducement to females to come closer, and the changing frequencies a means of calling attention to themselves. Tom Thompson has found that temporal or time differences in the call also can provide information to the whales. The length of a single sound, or the rhythm of the whole song, may contain information not only about an individual whale but also about his stock.

A fascinating aspect of humpback songs is that they exhibit oceanic, and perhaps regional, dialects (see Fig. 36). Although all humpback songs are similar, Pacific whales sing a different song from Atlantic whales. There are also song differences between the whales in the northern and southern hemispheres. This seemingly simple statement of fact is the result of a complex research project that was beset with problems. We shall describe this project in some detail by way of illustrating some of the little-known hazards of whale research.

It began in 1979 when a group from our laboratory and associates from the University of West Virginia and Shippensburg State College, Pennsylvania, decided to make a concentrated effort to document the existence of dialects. The plan was to send people to the Cape Verde Islands off Africa, the West Indies, the west coast of Mexico, Hawaii, and the northern Marianas. Six months later, during the breeding season in the southern hemisphere, a team would be sent to Tonga in the South Pacific. Songs would be simultaneously recorded in the north. Later we could analyze these songs and those recorded in Tonga and determine whether oceanic dialects existed. This simple-sounding project proved very hard to carry out.

First we had to obtain scientific permits to work in the various countries involved. We wanted to be sure everything was in order, since just months before we had been threatened with lawsuits and jail over a misunderstanding involving permits. Because obtaining permits to work in other countries through the State Department can take months, we decided to contact the countries involved directly.

When we inquired about a permit for working with whales in the Cape Verde Islands, government officials there became suspicious about our real purpose; only recently the Russians had proposed a submarine base in the islands in return for economic aid. The officials finally decided we were legitimate and granted the permit.

Acquiring a permit to work on Silver Bank, off the Dominican Republic, was also difficult. Because of the discovery of some $800 million worth of treasure on Silver Bank, the government was concerned that we might really be interested in the treasure. Sue Whiting and Flip Harrington, who were to run the 40-foot sailboat we had chartered, solved the problem by having a friend write to Senator McGovern, who somehow contacted the Dominican Republic. We eventually received a permit signed by the president of the Dominican Republic himself.

We went through the State Department to get our permit to work in Mexico, receiving a verbal permit over the phone but no written confirmation. Our United States permit was sufficient for working in the northern Marianas, which had just become our newest commonwealth territory. We were now ready to begin. One two-man team flew to São Tiago in the Cape Verdes; another

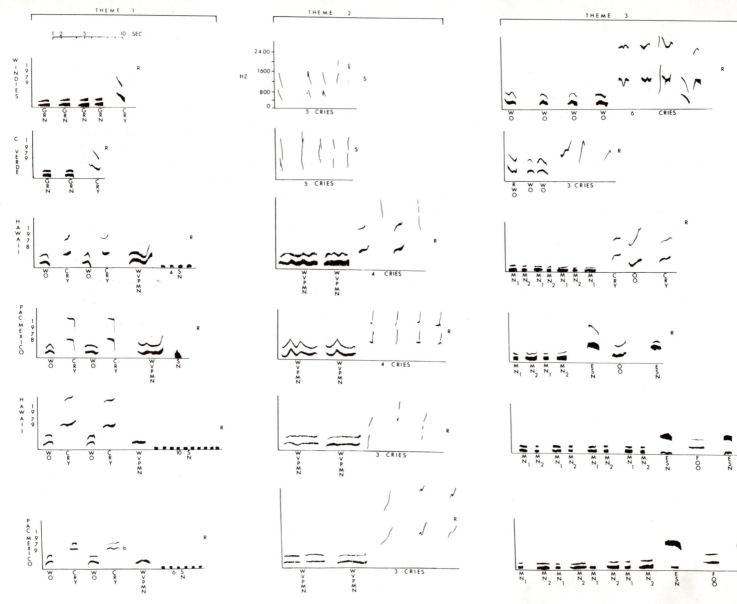

THEME 1 THEME 2 THEME 3

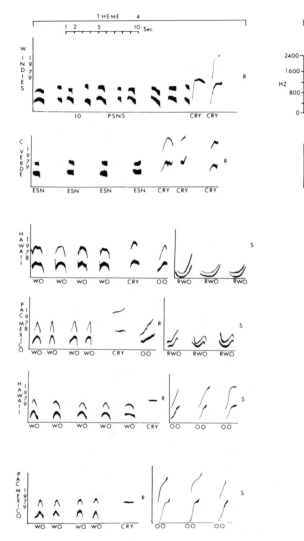

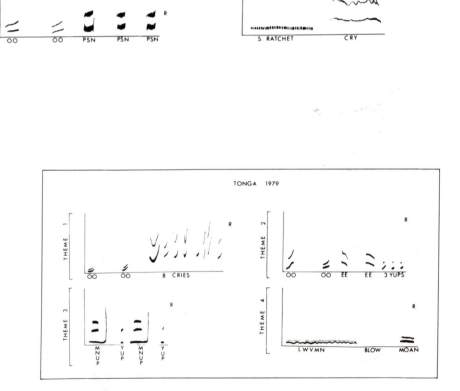

Fig. 36. Tracings of audio *spectrographs* of the 1978 hump-back song from Hawaii and Pacific Mexico, and the 1979 song from the West Indies, the Cape Verde Islands, Hawaii, Pacific Mexico, and Tonga. Each song is divided into themes. An R after a theme indicates that it is repeated; an S after a phrase indicates that it is a series of sounds. Letters under each sound are abbreviations or names given to the sound to facilitate analysis. Clearly there are geographic as well as year-to-year differences in the songs. *By Lois K. Winn.*

to Puerto Plata, Dominican Republic; a third to Puerto Vallarta, Mexico; and a fourth to Hawaii.

Bill Steiner and John Brayman managed to pick up an almost seaworthy boat in São Tiago. For over a week they traveled around the island but did not hear or see any whales. They then tried twice to get to Sal Island, where whales had been caught in the past, but heavy seas pushed them back each time. Once, while they were at sea, their engine broke down and they began to drift toward Brazil; they finally made it back to shore by using a rubber life raft as a sail. On their last day in the islands they decided to fly to Sal. They spent the day looking seaward without any sightings, but just before dusk a whale was spotted. They rushed to the village, hired a fisherman with a dory for a case of beer, and went out into the bay just as darkness fell. The hydrophone was lowered, and they were immediately rewarded with the sounds of a singing humpback. They recorded for over two hours that night and flew home the next morning.

Meanwhile Joe Marshall and Herb Hays arrived in the Dominican Republic. Everyone was on edge during the cruise because while the group was recording whales on Silver Bank, it was constantly followed by a Dominican gunboat. Even in port the researchers were watched and followed. Striking up a conversation with one of the surveillance men, Herb discovered that the man knew his life history and had somehow obtained a photograph of him taken ten years before!

When Tom Thompson and Bill Lawton arrived in Puerto Vallarta, the boat they were to meet was missing. They knew that its crew was uneasy about setting sail without a written permit because of the possibility of arrest or confiscation of the boat for illegal activities. Learning at last that the boat was in Acapulco, they left to meet it, only to learn when they arrived that the message had been mistranslated: the boat had left four hours earlier for Puerto Vallarta. They radioed from the yacht club at 3 A.M., and a Japanese freighter relayed the message to the boat, which returned to Acapulco. Eventually they had an uneventful cruise and made many recordings.

Sheldon Fisher and Howard flew into Hawaii, made a few recordings, and prepared to leave for the northern Marianas. Just before they left, they received a message that the captain of the boat they had chartered was ill and would be unable to sail for at least a week; they spent several days in ideal weather on shore in the Marianas, waiting for the captain to recover. They finally set sail, looking forward especially to mealtime, since the captain had hired a Japanese cook who had taken a week off from her restaurant to go on the cruise. Rainy weather soon became a storm with winds of 40 to 50 knots, and the cook promptly became seasick and took to her bunk, where she remained for the entire cruise. There was one bad moment when the captain slipped, fell, and almost lost consciousness; neither Howard or Sheldon knew the first thing about sailing a boat. After three days of stormy weather they returned to port without seeing or hearing a single whale. Later we learned that a fisherman had sighted two humpbacks the day after they left.

Six months later Tom and Nancy Thompson left for Tonga. They had to wait in Pago Pago for their luggage, which had been sent to Tahiti by mistake, and arrived in Tonga without a permit owing to the slowness of the mail. The owners of the charter boat would not put to sea without a permit, since another boat had recently been confiscated for illegally fishing in the king's waters. Finally, thanks to the help of Tongan fisheries officials, the case was taken up by the king's cabinet, and on August 14, after eight days of excellent weather, the Thompsons received the cabinet's authorization to carry out research work on whales in Tongan waters "for a period of 30 days from August 6th." For the next several days the weather was stormy, but eventually recordings were made.

By comparison, the long hours of song analysis were relatively calm and peaceful. The results showed clearly that oceanic dialects do exist, perhaps as a result of stock segregation (see Fig. 36). Recently computer analysis of certain sounds within the song (cries and wavers) has shown that differences between whales in separate breeding areas are partly area-related and partly individual; it would appear that each humpback has its own unique signature or voice pattern. Songs from the Mexico and Hawaii breeding populations were clearly different. These two groups of whales feed in the same areas of the North Pacific and are considered a single stock; thus the differences in their songs could represent sub-stock dialects. Recently, Hal Whitehead has recorded another song dialect in the northern Indian Ocean.

Although much has been learned about the humpback songs, many questions still remain. Why does the song change each year, do females ever sing, and what kinds of information are in the songs? Observations of the behavior of singing whales will help provide answers to some of these questions.

In recent years researchers in both Hawaii and the West Indies have been able to observe singing humpbacks for long periods. Peter Tyack has watched singers pursue nonsingers, and he has also seen them join a group, stop singing, and engage in behavior usually associated with courtship and mating. Others have seen singing humpbacks exhibiting aggressive behavior toward other humpbacks. All these behaviors imply that singing whales are males vying for the attention of females. Tyack also noted that as the breeding season progressed, the singers sang for longer periods.

Some whales may even continue to sing as they migrate to their feeding areas; we have recorded a few full songs off New England at the end of the migration period, as have D. McSweeney and W. Dolphin in Alaska. Perhaps unmated males are making one last desperate attempt to attract a female. Most singing whales, however, have stopped singing when they begin the migration to the north.

9. JOURNEY NORTH

During March some of the humpbacks in the West Indies begin to gather into groups and move toward the North Atlantic feeding grounds. Although most of the calves are growing fat and strong on their mothers' milk, many adults are in poor condition because they have eaten little or nothing since late fall. The journey to the north is perhaps the most difficult for the whales, and for some it will be their last.

Females with calves are usually among the last to leave the calving and mating grounds, and they must travel more slowly than the others. Often they will stop to idle or play, sometimes even changing direction temporarily but always moving toward the north at an average speed of 1.3 knots. Single animals and groups without calves move more rapidly, averaging from three to five knots.

Most of the whales take the same route to the north as they took on their journey south. Our data show that humpbacks on Silver, Mouchoir, and Navidad banks go outside the 100-fathom line along the United States to feeding areas off New England, Nova Scotia, Newfoundland, Greenland, and Iceland. Humpbacks pass New Zealand each year on their way to calving grounds off Tonga and other islands. In other areas of the world the routes are probably just as specific.

On one of our research cruises we took the same route home to New England as the humpbacks, hoping to encounter migrating whales along the way. The ocean between the West Indies and Bermuda is desolate. Vast patches of sargasso weed formed neat windows before us, as if we were in some great field of raked grass. For two days we traveled without sighting any life in the sea. On the third day, in the middle of the Sargasso Sea, a storm came up; dry and snug, we braced ourselves against the constant jolting of the ship. Suddenly in the swell ahead we glimpsed a black back. We hurried outside, hardly able to breathe as we were assaulted by water from above and below. As the ship plunged into a trough, the sea became smooth for a moment and two whales surfaced just a few feet from the hull, their blows hoarse and raspy. Between them a calf struggled upward, blew, and fell back with a splash. As suddenly as they appeared, the whales disappeared.

Whales can escape the fury of a storm by remaining in the tranquil waters beneath the surface, but evenutally they must come up for air. For a healthy adult whale, with its great strength and powerful muscles, a storm is easily managed. Storms are much harder on young, weak, or disabled animals, whose journey may never be completed.

Storms are not the only dangers whales face during

migrations. Sharks and killer whales are always ready to prey on the young and helpless. Another reported enemy of the humpback is the swordfish, but there is little scientific evidence for this assertion other than two reported instances of swords from swordfish being found embedded in the carcasses of whales. Swordfish have been known to charge boats and actually pierce the hulls, but as far as we know no one has ever witnessed such an attack on a whale.

Some early accounts have the swordfish and the thresher shark working together to overcome the whale. Two of these pertain to the humpbacks in Bermuda. In 1625 William Strachy wrote:

I forebeare to speake what a sorte of whales wee have seene hard aboard the shore followed sometime by the Swordfish and the Threasher, the sport whereof was not unpleasant. The sword-fish, with his sharpe and needle finne pricking him into the belly when he would sinke and fall into the sea; and when hee startled upward from his wounds, the threasher with his large Fins (like Flayles) beating him above water. The example whereof gives us (saith Oviedus) to understand, that in the self same perile and danger doe men live in this mortal life, wherein is no certaine security neither in high estate or low.

Another eyewitness account of the same time is similar:

Whales there are in great store at that time of the yeare when they come in, which time of their coming is lo Februarie and tarry till June. Likewise there commeth in two other fishes with them, but such as the whale had rather bee without their companie; one is called a Sword-fish, the other a Threasher;

the sword-fish swimmes under the whale and pricketh him upward; the Threasher keepeth about him, and with a mightye great thing like unto a flayle hee so bangeth the whale, that he will reare as though it thundered, and doth give him such blowes with his weapon that you would think it to be a crake of great shot.

It would be logical to attribute these accounts to the antics of the humpback. Someone seeing the humpback leap from the water for the first time may very well believe that the animal must have been attacked from below; if by some chance a swordfish happens to be in the vicinity, the observer is provided with a reasonable explanation. Unfortunately the solution is not that simple.

These stories persisted. The zoologist Glover Allen, author of *The Whalebone Whales of New England* (1916), believed that the flailings of the humpback were often mistaken for a great battle between sea monsters. Noting that such battles usually took place at the time when the humpbacks were passing on their migration from the south to the north, he reported that in his time early summer newspapers would often "include a vivid account of a terrific battle viewed by the astonished passengers of some incoming steamer." One such incident was reported in the *Nantucket Inquirer and Mirror* in June 1909:

A remarkable fight between monsters of the sea was witnessed by the passengers and crew of the steamer *Esparta*, which arrived at Boston from Port Limon, Costa Rica, on Monday.

The thrilling battle occurred south of Nantucket South Shoal lightship, between a whale and another great fish believed to be a swordfish. The whale was vanquished.

The whale was the only one of the two fighters visible to the passengers and crew. The great mammal lashed its tail violently, churning the waters into a mass of foam, while it was believed to be attacking the swordfish with its teeth. Several irregular plunges appeared to indicate a successful plunge by the fish beneath and finally the great whale was seen to throw its massive bulk clear of the water and sink from sight.

Allen concluded from this account that the passengers were merely observing a humpback whale lobtailing, finally breaching, then continuing on its way. He did concede, however, that the whale might have been attacked by a killer whale.

Killer whales are notorious predators and do occasionally attack other whales. Historically sailors referred to them as threshers; to further complicate matters, they were also known as swordfish because of their high dorsal fins. Killers are toothed whales about 30 feet long. They travel in schools sometimes numbering over a hundred, and hunt in packs of three to six. When they attack another whale they sometimes force its mouth open and tear out the tongue, which seems to be their favorite delicacy.

Perhaps the earliest account of killer whale attacks on whales was written by Pliny. He told of the great whales coming to the bays on the coast of Spain to breed after midwinter. Here the killer whales hunt them, "and

deadly enemies they bee unto the foresaid whales." When the victims attempted to save themselves by flight, the killers cut them off, killing them in the narrow seas or chasing them up on shore.

In the 1950's Chittleborough witnessed a killer whale attack during reconnaissance flights over the whaling stations off the western Australian coast. Two humpbacks and a calf were slowly making their way south to the feeding grounds when a group of four or five killer whales attacked them. One of the humpbacks, probably the mother, swam closer to the calf as if to protect it. The other humpback faced the killers and charged them, using its tail to attack. The killer whales hesitated, then finally gave up, leaving the family of humpbacks to resume their journey.

Although killer whales are carnivores, they do not always attack whales; indeed they have frequently been sighted feeding peacefully among humpbacks. Their predatory behavior is most likely affected by the food that is available to them; if sufficient fish are available, they leave other whales alone. Yet one unique shore fishery in Australia actually used killer whales to capture humpbacks.

The fishery was opened by Alexander Davidson in 1866 in Twofold Bay, in the southeastern corner of Australia, which lies along the humpback's route to the tropics. In July the killer whales would arrive and deploy themselves at various positions outside the bay, waiting, the whalemen believed, for migrating humpbacks to ap-

pear. When one approached, the killer whales would become excited, slapping the water with their tails and
making a great deal of noise that alerted the whalers to
their presence.

The killers would then harass the humpback, forcing
it into the calmer waters of the bay, where the whalers
had a much easier time capturing it. With this "assistance" by killer whales a humpback could be captured
within an hour; without such help it often took over
twelve hours to make a kill.

W. J. Dakin, in his book *Whalemen Adventurers*
(1934), described a typical hunt at this fishery: "Four of
the killers separate from the rest and whilst two of them
station themselves underneath the head of the whale, so
preventing her from sounding, the others swim side by
side, from time to time throwing themselves out of the
water on top of her and right across her blowhole." They
were speedily thrown off again, but the action was continued, as if the killers were well aware that their movements critically hindered the breathing of the whale.
After the humpback had been lanced and killed, the
whalers would attach an anchor to the harpoon line,
drop it, and return to their homes. The killer whales
would press the jaws of the dead whale up and down until they managed to get the mouth open. Once successful, they would dive in for their share of the tongue,
then, taking hold of the head, the tail flukes, and the
fins, they could pull the whale beneath the water (see
Fig. 37). After 20 to 28 hours, the whale would rise

Fig. 37. This stranded humpback was at some point attacked
by a killer whale, though the attack did not cause its death.
The flukes clearly show scars from the killer's teeth.

again, distended by gases, and the killer whales would interfere no more. The whalemen would then tow the carcass ashore for processing.

The leader of the killer whale pack, known as Old Tom, often alerted the whalers when the pack had surrounded a whale that could not be seen by the lookout in the tower. He would slap the water with his tail in front of the tower until the whalemen manned their boats and followed him to the unfortunate whale. Old Tom also had an annoying habit of grabbing the harpoon line in his teeth and hanging on as if he enjoyed being dragged through the water. Sometimes he would hang on for twenty or thirty minutes at a time.

On several occasions whaleboats were capsized and whalers found themselves in the water with killer whales. They were apprehensive at first but soon learned that the killers would not harm them. This relationship between man and killer whale was maintained by Alexander Davidson, who began the fishery, and later by his sons John and George, until 1932, when Old Tom died of old age.

To judge from aboriginal folklore, killer whales were encountered visiting Twofold Bay long before the Davidsons began their operation. The aborigines of the area regarded the whales as incarnate spirits of their own departed ancestors, and in this belief went so far as to give names to individual killers. If, as seems likely, they made use of the whale carcasses discarded by the killers, these welcome "gifts" no doubt reinforced their convictions.

Although many alleged attacks on humpbacks by swordfish and thresher sharks must in fact have been attacks by killer whales, the possibility of thresher shark attacks cannot be simply dismissed (see Fig. 38). Fishermen in Freeport, Nova Scotia, assert that they have repeatedly witnessed attacks on large whales by swordfish and thresher sharks. The thresher was described variously as chasing the whale or attacking it on the surface, the swordfish as aiding the shark by jabbing at the whale's belly from below and preventing it from diving. Interestingly enough none of the fishermen ever actually saw the swordfish, but they were convinced that it was the swordfish that prevented the whale from diving.

Scientific evidence is sparse concerning thresher attacks on whales, although the naturalist Frank Bullen wrote plausibly in 1904 of witnessing such an attack:

Here I am aware that I am upon highly controversial grounds, since very eminent professors of natural history deny that the thresher does attack the whale. They say, with warrant I fail to understand at all, that what the sailor has mistaken for the attack of the thresher on the whale has been the antics or gambols of the humpback whale, which has long arms (fifteen feet or so), and is fond of waving them in the air and bringing them down upon the water with a loud smack. They are entirely wrong.

Also, I have seen the thresher shark attacking the whale at close quarters, so close indeed that every movement of the shark and his victim was plainly visible, and I can hardly imagine any one mistaking the gambols of the whale for this curious attack. The shark appears to balance himself upon his head in the water, with the whole of his enormous flail-like

Fig. 38. This whale with its pectoral fin upraised and the tip of its fluke showing illustrates how tales of battles between humpbacks and the threshers may have originated. *Photo by Jeffrey Goodyear.*

Fig. 39. Jim Hain and some of his crew approach the bloated carcass of a dead humpback found floating at sea.

flukes in the air at the moment of striking; then, when the blow has been delivered there is a quick descent and return, like the lashing of a gigantic whip, while the blows are audible for two miles on a calm day. So heavy are they that strips of blubber are cut by them from the back of the hapless whale four to six inches wide, and two to five feet in length.

It is most unlikely that a shark would attack a strong, healthy whale. Its prey would more likely be a calf or a sick adult already weakened by disease or parasites. James Hain, once chief scientist aboard the R/V *Westward*, told us of an encounter he had while cruising off the coast of Maine. A whale was spotted floating belly up, and the crew launched a rubber raft and went over to examine the animal (see Fig. 39). It was a female humpback, grotesquely bloated by the gases of decomposition. Measurements were taken, along with detailed notes on scars and wounds. As it began to get dark, several crew members in the raft reported seeing a fin cutting the water nearby. Jim continued to work on the dead whale, assuming that this crew was just nervous because it was their first time at sea.

As he was taking a skin sample, a shark suddenly leaped from the water and took a bite from the side of the dead whale. Jim estimated the shark's length at 10 to

12 feet and thought that it might have been a thresher. They did not linger to make a positive identification, but returned to the ship for the night after securing the whale to the ship. The next morning, after checking for sharks, Jim and several of his crew dived under the whale to examine the underside, where they found numerous shark bites on the belly, head, and flukes. The cause of this whale's death was never determined.

Mortality, whether from predators or disease, is probably at its highest during the times of migration. We say "probably" because scientists know very little about the causes of death in whales. Whales that die at sea sink and are disposed of naturally; encounters with dead whales at sea are relatively rare. Occasionally a dead whale will wash ashore, but most such whales are young (indicating how critical are the first few years of a whale's life).

In early history the carcass of a whale was regarded as a gift from the sea, since it represented food, light, and heat. Later, especially in England, beached whales became the property of the king and queen, who divided the spoils; the king received the head, the queen the tail. Exactly what they did with these parts is not clear.

When shore whaling began in New England, many more whales were struck than killed, and eventually animals who had died of their wounds would wash on shore. Many arguments resulted—local Indians, the whalers, and the person who discovered the carcass would all claim ownership—and laws had to be enacted to govern the disposal of stranded whales. According to records of the Plymouth and Massachusetts Bay Colonies Session of the General Court of August 8, 1645, the auditor general was responsible for dealing with waifs, strays, goods lost, shipwrecks, and stranded whales.

Until a few years ago, very little scientific interest was shown when a whale stranded. People's major concern was disposing of the carcass before it became offensive. Rarely was a scientist notified and given the opportunity to examine the dead whale in relative solitude. Today, however, this has changed. With a number of whales on the endangered species list, stranded whales usually get much publicity. The recent enactment of the Marine Mammal Protection Act in the United States makes it a crime to touch or interfere with any whales, dead or alive, without an official permit. Although it is easy enough for scientists to obtain such a permit, it may not be easy for them to work in the presence of a large crowd of onlookers (see Fig. 40).

In the past few years much has been written about cruel scientists rushing in to dissect a poor whale before it has even breathed its last. Some groups insist that dead whales should not be violated in any way and have evoked considerable sympathy from the public. James Mead, a curator of the Smithsonian Institution, was notified in November 1973 that a humpback whale had stranded for five days on a North Carolina beach. When Jim and his assistants arrived it was still alive, though its skin was sunburned and peeling, the baleen was torn off on

Fig. 40. Scientists examine a stranded humpback as people gather to watch.

one side, someone had hacked a hole in its skin just behind the blowhole, there was a ship's marlin spike (used for splicing rope) protruding from the head, and it had also been shot several times with a 38-caliber revolver. A rope around the tail suggested that the whale might have been towed or perhaps caught in a net and injured. Jim contacted the National Marine Fisheries Service and obtained permission to put the whale out of its misery, but could not find a sufficient quantity of drugs to accomplish the task. After taking a small blood sample from a vein in the fluke, Jim and his assistant tried to ease the whale's suffering by wetting its skin and then sat down with the whale to wait. Sitting throughout the cold,

windy night they often heard sounds coming from the dying whale. In the morning it was dead. No animal in its condition could have survived.

As they proceeded with their next job, to determine the cause of death, they were informed that if they had attempted to kill this whale they would be faced with a lawsuit. A protectionist group had falsely accused them of taking skin samples from the whale while it was alive, and in general of having no regard for the whale's life.

Incidents like this only make a scientist's job more difficult. Scientists examining dead whales must work in the open, often surrounded by untrained, emotional people; by contrast, and for good reasons, autopsies on humans are not done in public. The necropsy on this particular humpback showed that its lungs were filled with fluid, the result of pneumonia, and its liver was destroyed. By examining dead whales we add to our knowledge of the conditions that cause their death, and thus to the likelihood of saving stranded whales in the future. Some argue that enough dead whales were examined during the days of whaling; but such whales were on the whole healthy animals, and trained veterinarians did not participate in the studies.

In some cases all we know about certain rare species is what we can learn from a stranded individual. Scientists first identify the species and sex of the animal. Detailed body measurements are then made, not only of total body length but of every body part and its relationship to other parts. Samples of parasites are taken from all body orifices and skin folds. The skin is examined for wounds, scars, and scratches, and the animal is photographed. Skin and blubber samples are taken for pesticide analysis. Finally the internal organs are examined for parasites and pathological conditions. If the animal is a female, the ovaries are studied to determine its relative age. From these studies we have learned that when a large whale strands today, it is usually suffering from a fatal condition and cannot be saved.

Necropsies have shown that whales suffer from many of the same diseases that afflict man: hypertension, meningitis, and other neurological diseases; cirrhosis of the liver; and pneumonia. Internal parasites are another problem. Humpbacks and other rorquals are often infected with a nematode, *Crassicauda crassicauda*, which infests the kidneys and the urethra; yet if a whale can survive infancy, an equilibrium is reached between this parasite and its host. From stranding studies we learn not only about diseases but also about migration patterns, the presence of pollutants, and other changes that may be occurring within a population. The more we know about a species, not only in life but also in death, the greater our chances of ensuring its survival.

Fortunately for the humpback, predation and disease are relatively rare, and for most of the herd the journey north is uneventful. Large numbers of humpbacks are seen around Bermuda from March to May, suggesting that they are animals from the West Indies en route to the feeding grounds. Alternatively some Bermuda humpbacks may be a distinct breeding population; we do not yet know for sure. According to Greg Stone and Steven Katona, Bermuda humpbacks have been seen in the Gulf of Maine, Labrador and Newfoundland, and near Greenland, as well as in three areas of the West Indies.

The early settlers of Bermuda found humpbacks in great numbers. The first account was written by Sylvanus Jourdan in 1610: "There are also great plentie of whales which I conceive are very easie to be killed for they come so usually and ordinarillie to the shore that wee heard them oftentimes in the night abed and have seen many of them neare the shoare in the day time." Soon after this the settlers established a shore fishery. Evidently they had some difficulty at first; the whales were described as "quick and lively," and when struck in deep water, they dived so violently that a boat could be pulled under if the rope was not cut. We suspect that the whales in question were male humpbacks, known for their wildness, who usually stayed in the deeper water outside the reefs. According to Richard Stafford, who wrote in 1685,

We have heareabouts very many sorts of fishes. There is amongst them great store of whales, which in March, April and May, use our coast. I myself killed many of them. Their females have abundance of milk, which the young ones suck out of the teats that grow by their navel. They have no teeth, but feed on moss growing on the rocks at the bottom during these three months, and at no other season of the year. When it is consumed and gone, the whales go away also. These we kill for their oil.

The meat of the whale was consumed by poor families who could not afford other meat and poultry; calf meat was particularly desired. By the nineteenth century the population of humpbacks frequenting the shores of Bermuda had been greatly reduced both by local fisheries and by whalers operating to the north and south.

The late Frank Watlington of the Bermuda SOFAR station listened to and recorded humpbacks for many years. In 1976 he heard very few whales, and he told us that every year he was hearing fewer. Our data and those of other scientists support this contention. A historic native fishery operated by the Eskimos of Greenland may be responsible; though the fishery is primarily for Minke whales, humpbacks are also taken.

Proper management of stocks is beneficial to both the hunter and the whale, but whale hunting is a very emotional issue. To better understand some of the problems involved, we must also understand the long history of man the hunter.

10. MAN THE HUNTER

Humans learned to make use of whale products at an early date. Whales were probably more abundant before large-scale hunting, and strandings were quite frequent. People dwelling along the coast would make use of the entire whale carcass: the oil was used for light and heat, the bones were fashioned into implements, the intestines were twisted into twine, and the meat was eaten. Whale meat is red and resembles beef in texture and flavor; richer in protein than pork or beef, it has no parasites that are transferrable to man.

Food-gathering cultures gave way to hunting cultures, men learned to hunt porpoise from small canoes, growing more skilled until they were eventually taking the larger whales. Some Eskimos living along the coasts of Alaska still hunt whales from small boats using hand harpoons, though most now use bomb lances, which contain an explosive charge that kills the animal almost instantly.

Although we know very little about early whaling, we can assume that people did not venture very far from land and that their prey must accordingly have been whales that inhabited areas fairly close to shore. Since most humpback stocks feed near shore, humpbacks must be one of the earliest species hunted.

The first recorded whaling industry began along the coast of Persia (Iran) about 2000 B.C. Whales were later hunted in the Mediterranean by the Phoenicians, a Semitic tribe, who were skilled shipbuilders and unrivaled in maritime commerce for many centuries. Whales (humpbacks were in these waters) were processed at shore stations, and the oil, a valuable commodity, was sold. At one time it was thought that the Phoenicians restricted their voyages to the Mediterranean, but recent interpretations suggest that they may have traveled as far as the Caribbean and perhaps to the eastern shore of the North American continent. Whether whale hunting took place on these particular voyages is not known.

Another great whaling culture began around 890 A.D. The Norse, or Vikings, hunted whales regularly in the North Atlantic; their voyages may have taken them as far south as present-day New England. Captured whales were brought to shore and processed on the spot. Norwegians today continue to hunt whales for food.

During the eleventh century the Basques, living on the Atlantic coasts of Spain and France, maintained an extensive whaling industry. Shore whalers at first, they posted lookouts along the coast and set out in small boats when a whale was sighted; after a whale was killed it was towed to shore for processing. Basque whalers, who contributed heavily to the destruction of right whale stocks, probably also took humpbacks in the eastern North Atlantic. When whales became relatively scarce in these waters in the thirteenth century, the Basque whalers extended their operations throughout the North Atlantic.

We know a little more about the fate of the whale stocks in the western North Atlantic. The first European explorers were impressed with the number of whales they encountered in these waters. In 1578 Frobisher, voyaging to Davis Strait (between Greenland and Baffin Island), noted, "Wee meete with manye great whales." In Greenland waters, John Davis wrote in 1585, "Great numbers of whales were seen." In 1610, also off Greenland, Henry Hudson "saw store of whales," and in 1612 William Baffin "saw many whales."

According to an 1872 account by P. Fischer, Basque whalers "in 1372 reached the banks of Newfoundland where they observed whales in abundance." Although this account has little evidence to support it, the Basques were probably the first to carry on a whale fishery in these waters. The first recorded voyage of the Basques to North America took place in 1538–1540. By 1578, Anthony Parkhurst reported in a letter to Hakluyt that from 20 to 30 Basque whaling vessels were operating off Newfoundland.

Earlier whalers were not interested in colonization, but the abundance of whales and fish off Cape Cod was responsible for the settlement of that coast by the English. In 1605 George Waymouth remarked on the "great store" of codfish and whales near Nantucket, and described how Indians hunted whales. The Plymouth colonists William Bradford and Edward Winslow wrote in their journal of 1620, "Every day we saw whales playing hard by us." Some experienced whalers in their crew thought Cape Cod would be better than Greenland for whaling and planned to establish a fishery there the following year. William Morrel, who visited Plymouth in 1623, published a poem on his return to England that included these lines:

> The mighty whale doth in these harbors lye,
> Whose oyl the careful merchant deare will buy.

We have no records of the earliest shore fisheries. The first indication we have that the colonists were engaged in whaling dates from 1650, when many disputes between rival whalers over ownership of dead whales are recorded. The same legal records contain information on laws governing drift whales. They also state that much oil was being shipped to London, and that the crown was taking one barrel of oil for every whale caught.

Most shore fisheries operated in the same way. Lookouts were posted along the coast, and when a whale was sighted the whale was first harpooned and later killed by hand lances; the carcass was towed to shore and processed on the beach. The blubber was stripped from the carcass and boiled down to oil, and the remains were left on the beach to be carried away by the tides. For some reason, even in times of near starvation, the early settlers did not eat whale meat.

Since many of the humpback's feeding areas were relatively close to shore, this whale was particularly vulnerable to the hunter. The earliest account of the capture of a humpback was in 1608, when a party of Indians killed a stranded animal on Nantucket. This may have been the beginning of the Nantucket shore whale fishery, which attained its greatest prosperity in 1726 with the taking of 86 whales. The records do not state what species of whales were killed, but they probably were right whales and humpbacks. By 1748 far fewer whales were seen feeding off Cape Cod. According to J. B. Felt in the *Annals of Salem* for 1748: "Whales formerly for many successive years set in along-side by Cape Cod. After some years they left this ground and passed farther off upon the banks at some distance from the shore. At present the whales take their course in deep water whereupon a pace our whalers design to follow them."

After the Revolutionary War and until the War of 1812, New England whalers continued to take humpbacks on the shoals eastward of Nantucket, where, according to Obed Macy's *History of Nantucket* (1835), "these as well as cod fish were plenty." Although the people had no use for the meat, they apparently enjoyed the crisp bits of fat that were left after boiling down the blubber. In 1875 a writer in the *Nantucket Inquirer* recalled a memory from his childhood: "We were often made glad by the arrival of a fortunate 'humpbacker' for the crisp bits of 'flukes and scraps' resulting from the trying out of blubber on the shore." The introduction of the bomb lance in the mid-1800's aided in the taking of the humpback, and many were killed in the next two decades.

Almost two hundred years of continuous hunting finally began to take its toll. Twenty humpbacks were taken in Provincetown Harbor in May 1881, considered an unusually large number at that time. The last humpback to be taken in Massachusetts coastal waters was killed in 1895; after that shore whaling was abandoned.

While the colonists were engaged in shore whaling, and after the peace of 1763, a group of New England Loyalists who were also experienced whalemen settled on the shores of the Gaspé in the Gulf of St. Lawrence. They began with small craft, but in a short time they had twelve whaling schooners. The main species hunted was the humpback. They and their successors prospered for over a century, but by 1875 only three schooners were operating out of the port of Gaspé. The reason for this decline is given in the 1875 *Report of the Commissioner of the Fisheries of Canada*: "Whales had been so eagerly pursued for some years past by Gaspé fishermen that they disappeared for the same causes, I presume, which led them to abandon the shores of Europe and America. This fishery having been unremunerative was abandoned."

Although humpbacks may have deserted some areas

because of the changing movements of their food supply, shore whaling unquestionably had a significant impact on their numbers in the North Atlantic. Whalers' journeys eventually brought them to the humpbacks' tropical mating and calving grounds, where they found them concentrated in large groups in relatively small areas. In the western North Atlantic they were taken at Bermuda and in the West Indies from Haiti to the Venezuelan coast. Eastern North Atlantic humpbacks were captured off the coast of Spain and around the Canary and Cape Verde Islands. In the South Atlantic both foreign whalers and natives took humpbacks on the mating and calving grounds off South Africa, Angola and the French Congo, and in mid-ocean around St. Helena Island.

F. D. Bennett, in his book *Narrative of a Whaling Voyage . . . 1833–1836*, reported that humpbacks were frequently seen in the deep waters around St. Helena, where local fisherman believed they destroyed fish and frightened them away from the coast. The flesh of a humpback calf was considered a delicacy by every resident of the island, from the governor to the lowliest slave. Fishermen on St. Helena report that fewer humpbacks have visited this island since World War II.

When the Americans reached California they found practically untouched stocks of humpbacks regularly migrating past the coast. At Monterey a shore whaling station was established in 1851 to take the California gray whale and the humpback. Since both species annually pass close to the coast on their migrations to the southern breeding grounds, they were easily hunted from shore. Soon other stations arose, until there were eleven companies operating from Half Moon Bay, south of San Francisco, to Point Abanda in Baja California. By the late nineteenth century, however, shore whaling in California was rapidly declining; whales were scarce, and those that were seen were difficult to catch. One more attempt at shore whaling was made in 1939 when a station at Field's Landing, near Eureka, California, went into operation and took a number of fin, sperm, and humpback whales; but this fishery soon closed, marking the demise of shore whaling on the California coast. Thomas Dohl has noted an encouraging increase in humpback sightings off central California, starting with a low in 1980 of 87 individuals and rising to more than 300 in 1983.

Humpbacks were also pursued by the Indians of Vancouver and the Queen Charlotte Islands. Further to the north, Eskimos took humpbacks in the bays and around the islands of Alaska and in the Aleutian Islands. Eskimos pursued their prey in 30-foot wooden canoes, using harpoons made of ivory with a slate-stone point. The crew consisted of eight people, and some crews included women. After the whale had been harpooned, it was lanced with lances attached by rope to an inflated sealskin buoy. Eventually, when the whale died of its wounds, the buoy would mark the location of the carcass, which was then towed ashore and cut up. Like most primitive whale hunters, the Eskimos utilized the entire carcass. The blubber and flesh were eaten; the sinews were fashioned into ropes, cords, and bowstrings; the stomach and intestines were dried, inflated, and used to store oil. The oil was used as a substitute for butter or traded to neighboring tribes or occasionally to white traders. The natives had no use for the whalebone except to trade it to the foreign whalers who sometimes visited them.

Shore whaling for humpbacks has been carried out in almost every part of the world. Some were historic native fisheries, others were started by whalers who had given up deepwater whaling and settled in areas where humpbacks were still numerous. In the South Pacific humpbacks were taken off New Guinea, Australia, New Zealand, Fiji, and Tonga, and off the coasts of Ecuador and Colombia from Guayaquil to the Bay of Panama. In the eastern North Pacific humpbacks were taken in Hawaiian waters, off the California coast from San Francisco south, and in bays along the west coast of Mexico. In the western North Pacific shore operations were carried out in Japan, Korea, and the Marianas.

The first American whaling ship entered Hawaiian waters in 1818 in search of the sperm whale. The logbooks of the earliest whalers in Hawaiian waters make no mention of humpbacks, which seems unusual if they occurred there in any number; and Louis Herman, after extensive research, has concluded that the humpback may be a relative newcomer to the shores of Hawaii. Hawaiian legends, myths, ceremonies, rituals, and historical writings yield no evidence that humpback whales were known to pre-nineteenth-century Hawaiians. The first clear indication of their presence in Hawaiian waters comes in the 1850's in reports of shore whaling companies specifically formed to take humpbacks. These companies' shore operations, of which there were at least five, were operated strictly to obtain whale oil, which at this time was beginning to decline in economic value because of the surge in petroleum production. For this reason, and because stations probably greatly reduced the humpback's numbers, humpback whaling in Hawaiian waters was essentially over by the 1860's.

Shore whaling in Japan dates to the Stone Age. From that time to the present Japan has depended heavily on the whale for its supply of meat. Since a large part of the country is mountainous, and since every inch of the coastal plains is used for crops, the Japanese do not have the extensive grazing areas found in other countries. Dairy products are a luxury in Japan, and the people must look to the sea for their protein.

In 1674 a unique method of capturing fin and humpback whales was developed in Kishu. Before this time Japanese shore whalers used the same methods as the early New Englanders. The new method, like the old, employed lookouts posted on high areas along the coast.

When a whale was spotted small boats were launched, some of which headed out to sea, then turned and drove the whale toward the shore. When the whale reached shallow water, other whalers closed in and threw a net over it; the animal, entangled in the net, could now be harpooned. At the moment of death the whalers would chant three times, "May his soul rest in peace." Whales were towed ashore for processing. Since the nets were expensive, this fishery was financed and organized by the wealthy. Japanese net whaling was finally discontinued in 1909.

Another kind of net fishery was established in 1890 on North Island, New Zealand, by H. F. Cook, who had served for a time on a Yankee sperm whaler. Cook and his men attached a net to a cable that was strung over a channel through which humpbacks regularly passed; when a whale became entangled in the net, the whalers would kill it. At first rope nets were used, but later wire nets were found to be much more effective. The catches were small, the largest being nineteen humpbacks in one season. In 1910, when a steam catcher boat was built, this net fishery was abandoned.

As the numbers of humpbacks and other whales along the coasts of North America began to dwindle, whalers were forced to travel farther out to sea. At first they concentrated on bowhead and right whales, which yielded more oil than rorquals and did not sink when killed, an important consideration when whaling at sea. But soon these species became depleted from overfishing, and whalers turned to the more numerous rorquals.

Humpbacks were easier to take than other rorquals because of their habit of concentrating in large numbers on their mating grounds, where they were relatively unmolested except by native shore fisheries. Some New England whalers combined sperm and humpback whaling. Sailing in mid- to late January, they went directly to the humpbacks' mating grounds in the West Indies. In late April, after most of the humpbacks had returned to the north, the whalers headed for the Charlestown or Hatteras sperm whale grounds, returning to New England in September. Before long other calving areas were discovered along the coasts of South America and Africa, and around the islands of Cape Verde and St. Helena.

Whalers went to the Crozet and Desolation Islands primarily for sea elephants and right whales, but also found humpbacks in abundance. Soon humpbacks were also taken around Madagascar and in other parts of the Indian Ocean.

How easily a humpback was captured probably depended on its sex. Females with calves were easily taken, since the mother humpback displayed a great devotion to her youngster and would not leave it even if it had been wounded; knowing this, whalers generally attacked the calf first. Bull humpbacks were quite another matter. Whalers usually avoided them, since they often succeeded in smashing the boats of their pursuers. A whaler who operated a shore fishery on Barbados observed that whenever his men saw a cow and a calf they generally succeeded in taking them, but that when the bull was present their chances were poor. Happily for the whalers the bull was often not present, since bulls typically stayed in deep water while their mates calved in shallower, calmer areas nearer shore and remained there while their calves were very young.

Occasionally whalers would attempt to take lone bulls. Nordhoff was told of an easy method of capturing bull humpbacks, which was to sneak up on the whale and then create a great deal of noise by yelling, beating pans together, and blowing foghorns, thus terrifying the whale and rendering it unable to move. Nordhoff set out to test the method. He sighted a pair of humpbacks and approached them slowly; as soon as his boat was close enough, the crew began to make a racket in the manner decribed. One whale escaped, but the other "became in a manner paralized from fright, and lay still upon the water." Quickly they struck the motionless whale, but "the first touch of the iron gave back to him all his powers and quick as a flash an immense pair of flukes came down on the boat's bow, cutting off about three feet of it nearly as smoothly as though it had been sawed off." Apparently Nordhoff did not use this method again.

As we have seen, humpbacks do not yield as much oil as other whales. There are occasional reports of a humpback yielding 75 to 80 barrels, but these were probably unusually fat females. On the other hand, the coarse, black baleen of the humpback was especially suited for women's corsets, an important consideration at a time when all fashionable women wore this garment.

By the end of the nineteenth century whalers had moved into the Antarctic and begun for the first time to take humpbacks on their southern feeding grounds. By 1937 humpbacks were becoming scarce on some feeding grounds in the Antarctic and that year a minimum length of 35 feet was set by the First International Whaling Commission. This did little to ensure survival of the species, since more females than males exceeded the minimum, and in 1939 the Whaling Commission prohibited the killing of all humpbacks in the Antarctic. In 1946 the Second International Whaling Commission was formed, and in 1949 it authorized the taking of 1,245 humpbacks in the Antarctic each season. Later this was changed to a limit of four humpback whaling days each season, and still later, in 1963, humpback whaling was totally prohibited throughout the southern hemisphere.

Although humpback whales are protected throughout the world today and are considered an endangered species, several historic native fisheries continue to take a few each year. In the Pacific a small fishery was operated until recently by the natives of Tonga, an island group in the South Pacific. In the Atlantic there are humpback fisheries at Bequia, a small island in the Grenadines, at several native villages in Greenland, and in the Cape Verde Islands.

We visited the Bequia whaling station in February 1970. Leaving the lush tropical shores of St. Vincent Island to the north, we sailed to the brown, almost desertlike shores of Bequia, several miles away. A jeep and driver were waiting on shore to take us to the southeast coast, where the last of the whaling operations were carried out. Most of the island is sloped, with a steep rocky shoreline. Unlike nearby St. Vincent, Bequia has erratic rainfall, and there are often long periods of drought from which crops and livestock suffer. These conditions were directly responsible for the establishment of the whale fishery on Bequia.

From 1867 to 1870, when American whalers were making regular visits to the West Indies in search of sperm and humpback whales, over 250,000 gallons of whale oil was shipped from St. Vincent. The whalers, working close to shore, gave the natives ample opportunity to observe the chase, kill, and butchering; indeed, since it was difficult to obtain crews for whaling voyages, many native West Indians were signed on. One of these men was William Wallace, a native of Bequia of Scottish ancestry. Wallace began whaling around 1875, hoping to generate a new source of income for the people of Bequia. He began with three boats and a small shore station for processing. Soon afterward Joseph Olliverre started a whaling operation on Petit Nevis, an islet south of Bequia. From Bequia whaling spread to other islands in the St. Vincent Grenadines, and between 1893 and 1903 native fisheries transported 25,000 gallons of whale oil to St. Vincent for sale. By the 1920's humpbacks were becoming scarce, and it was difficult to capture even one in a season. Most of the native fisheries went bankrupt, and only one station, at Friendship Bay, Bequia, was still operating at the time of our visit.

As our jeep crested a hill, we could see the shores of Friendship Bay below us. The whaleboats were anchored, bobbing gently in the slight breeze. Most of the whalemen lived here or at Paget Farm, a small village further along. As we approached the village we could just glimpse the boiling pots and drying racks of the processing station on Petit Nevis (see Fig. 41). This station had been built in 1961, on the site of the earlier fishery constructed by Joseph Olliverre.

Athneal Olliverre, the head harpooner of the fishery, told us that the whaleboats there have several owners, some of them descendants of the original whalers. All members of the crew receive an equal share, regardless of who strikes the whale. At that time only two boats were operating, each with a crew of six. Hand harpoons were used to lance the whale, but a gun was occasionally used for the kill. The whaling season on Bequia runs from February to May, with the best months being February and March; the first humpback of the season had been taken a week before our visit. The two boats stay together and keep in contact with walkie-talkies. The calf is struck first but not killed, so that the mother will stay nearby. The bull is considered dangerous, and some encounters with bulls have resulted in damaged boats.

Each member of the crew receives a share of the meat, which is then distributed among about 400 people on the island. The rest is shipped to the market in Kingstown, St. Vincent, where it is in great demand (see Fig. 42). At that time the meat sold for 50 cents a pound,

Fig. 41. A female humpback is butchered at the processing station on Petit Nevis, just south of Bequia in the West Indies. *Photo by Jane Beck.*

Fig. 42. Humpback whale meat being sold in the market in St. Vincent in the West Indies. Each portion of meat is topped off with a piece of blubber and both are sun-dried.

making the value of one whale about four to five thousand British West Indian dollars. The bones are sold to the people, and only the baleen is discarded. Some of the oil is stored by the crew for later sale, the rest bottled and sold in the market. Considered a cure-all, the oil is used to ease strains, prevent colds, stop headaches, and promote a long life.

One heritage the whaling industry has left to the people of Bequia is their boat-building skills. Their whaleboats are modeled after the Nantucket whaleboat: they are double-ended with a V-shaped hull, are about 26 feet long, and carry a foresail and a mainsail. Although the whalers use their whaleboats only for whaling, similar boats have been built for fishing.

The whalers of Bequia are admired for their bravery and skill, and the heritage of boat-building, sailing, and whaling skills has contributed greatly to the culture of this small island. Since 1970 only one or two humpbacks a year have been caught, but people in Bequia still see whaling as important to the island's economy and have urged Olliverre to keep the fishery going. It will probably continue to exist as long as there are whales to hunt and a market for the meat. At least two humpbacks were killed in 1983.

Until 1978 the people of Tonga, in the South Pacific, operated a humpback fishery much like the one in Bequia. During the winter months July to October, whales were taken to the east of the Tongan group of islands, whose waters serve as the breeding grounds for a herd of humpbacks that summer in the Antarctic. Humpbacks were first hunted near Tonga by American and English whalers, one of whom, a man named Cook from a whaling family of Devon, England, remained to settle in Tonga. Tonga's humpback whale fishery centered around the Cook family, and the whalers remained a very close-knit society.

Whaling was carried out from the main island of Tongatapu and from Vava'u, 150 miles to the north. The men cruised in their 30-foot sailboats until a whale was sighted. Although hand harpoons were used, they were fitted with explosive heads that occasionally caused injuries to the whalers. Like the Bequians these men were skilled sailors and proud of their skills; they typically maneuvered to within ten or fifteen feet of the whale before harpooning it (see Fig. 43). Here the female was taken first and towed to shore for processing; the calf followed the carcass of the mother and was also taken later. These whalers rarely took bulls and usually avoided them. The demand for whale meat in Tonga was great; whenever a whale was landed, the shores were lined with people waiting to buy meat. An average of two or three whales were taken in Tonga per season, netting the whalers about one thousand Tongan dollars. In 1975 eight whales and one calf were taken. In that year Walter Cook, the senior member of the family, was fatally stricken with a heart attack after harpooning a whale.

Fig. 43. Using a crude hand harpoon a native of Tonga in the South Pacific strikes a humpback whale. *Photo by William Dawbin.*

Other family members continued to operate the fishery until 1978, when they gave up whaling as uneconomical.

Not much is known about the Greenland fishery. The native villages are small and widely scattered, and have little contact with the outside world during the winter months. Catches off Greenland are estimated to be as high as ten to twenty humpbacks each year. Current research by Hal Whitehead and by scientists from the Ocean Research Education Society in Gloucester, Massachusetts, suggests that the west Greenland humpbacks are a discrete substock of perhaps 200 to 300 animals. If this is true, catches as high as twenty a year could be depleting the population. At Cape Verde one or two humpbacks are taken every couple of years.

As far as we know, these are the only places in the world where humpback whales have been hunted since 1966. As long as such fisheries use primitive methods and the catch remains low, they do little damage to the stocks they are hunting, except possibly to the west Greenland substock. Since they are demonstrably important to the isolated economies of Greenland, Bequia, and Cape Verde, the exception made for them by the International Whaling Commission seems justified.

Though whaling more than any other single factor has historically been responsible for depleting the world population of humpbacks, the most critical problems facing the species today are very different. They relate to a potential shortage of food, to diseases caused by pollution, and to a general pressure to adapt caused by the rapidly increasing physical intrusion on the whale's territory by humans.

11. THE FUTURE

The survival of any species depends on its ability to adapt to changing environmental conditions. Humans have survived a variety of life-threatening conditions because their intelligence has allowed them to avoid or overcome threats to their survival. Evolution is a slow process. Conditions in the world today have changed rapidly because of our mismanagement of the environment, which has left little time for adaptation to occur. All living things face the threat of environmental damage so severe that escape becomes impossible.

Even without interference from humans conditions on earth are constantly changing. Some species have become so specialized that any slight change in environmental conditions would mean death to the entire population. Thus the blue whale has evolved into an extremely effective krill harvester because the fine, silky hairs of its baleen allow it to sift a maximum amount of microscopic organisms from the sea. As a result it feeds almost exclusivel on krill. If for some reason these small sea creatures were to disappear, the blue whale could not survive.

The humpback is perhaps the least specialized of the baleen whales. Its coarse baleen allows it to take not only krill but many kinds of fish and even mollusks. If one food supply becomes depleted, it can turn to another. But what if all food supplies were to become depleted? As the world's population increases, it will become more and more difficult to feed the hungry and people will inevitably seek more food from the oceans. Already it has become necessary to manage certain stocks of fish so that overfishing will not result in their destruction.

Krill itself has come in for this kind of attention; the staple diet of many whale species, it could someday rival soybeans as a source of protein for livestock and for humans. It is already being harvested by fishing fleets from Japan, the Soviet Union, and South Korea; the United States and other countries may soon follow. The potential annual harvest of krill is estimated to be more than 200 million tons. Needless to say, fish-eating whales face the same problem. Currently there are enough fish for both man and whale, but someday, whether because of ecological disaster or population pressure, the whale may become expendable. Although the humpback is more adaptable than most, the threat of starvation is real.

Another potential threat to whales is pollution. The oceans have become a vast dumping ground for various kinds of wastes. Pollution not only destroys many organisms, but causes damage that may become obvious only when it is too late. National policy toward conservation and environmental protection must reflect various human needs, and may accordingly change in the years ahead. In 1980 National Oceanic and Atmospheric Administration Chief Richard Frank argued that potential damage to marine environment may no longer be sufficient reason to delay drilling for offshore oil. Already oil drilling has begun in important feeding areas such as Georges Bank in the Atlantic, and oil exploration has been proposed for Silver Bank itself. No one really knows what effects an oil spill would have if it occurred in a critical whale habitat. Researchers are just beginning to understand the effects certain pollutants can have on living organisms.

Chlorinated hydrocarbons such as DDT, dieldren, and PCBs (polychlorinated biphenyls) are increasingly finding their way into the oceans. PCBs were used in a variety of ways prior to 1971, when it was discovered that they were potentially harmful when released into the environment. As a result the United States has drastically reduced the manufacturing of PCBs, but the Environmental Protection Agency estimates that we continue to import 500,000 pounds of this chemical each year. Although severe injury from short-term exposure to PCBs is not likely, long-term low-level exposure may cause a variety of problems. PCBs are difficult to eliminate from the environment, remain for long periods of time, and can be found throughout the world.

Until recently PCBs were dumped into the oceans, where they adhere to small particles in the water and are consumed by life forms throughout the food chain. In 1975 our laboratory analyzed tissue samples from dead humpbacks and other whales to determine the presence of chlorinated hydrocarbons. The humpbacks tested were found to have levels of PCBs ranging from 1.3 to 5.4 ppm (parts per million), and total DDT from 1.4 to 23.1 ppm. One old Atlantic pilot whale showed 114 ppm PCBs and 268 ppm DDT.

In 1968 five Japanese people died and a thousand others became ill after eating cooking oil contaminated with PCBs; some of these people had PCB levels as high as 75 ppm. When PCBs or other chlorinated hydrocarbons are injected in living organisms, they concentrate in fat tissue, the kidneys, the liver, the brain, muscles, and blood. Nursing females pass on high concentrations to their offspring through their milk.

Like DDT, PCBs can become quite concentrated in fish without doing any immediate damage. Since whales cannot avoid the contaminated food, they can eventually build up large concentrations of the chemicals in their bodies. In monkeys' diets containing concentrations as low as 2.5 ppm PCBs caused severe health problems such as loss of hair, acne lesions, and reduced rates of conception; offspring had PCB levels almost as high as their mothers. DDT and PCBs also cause other sublethal damage, notably decreased resistance to infection, alternation of hepatic enzyme activities, and liver damage.

There have been no direct studies of the effects of these chemicals on cetaceans, but several studies indicated that chemicals have caused reproductive failure in California sea lions and kit mortality in the mink. Reproductive failure would be especially difficult to detect in wild cetaceans, since birth occurs underwater and relatively little is known about the population levels of most species.

Some whales have also been found to have high levels of mercury in their systems. A bottlenose dolphin studied in Japan had a mercury level 150 times the legal limit for human consumption. Coastal species seem to be more affected, although sperm whales, a deep water species, were tested at over 2.3 ppm, nearly six times the acceptable level of 0.4 ppm. Mercury poisoning can cause birth defects, severe neurological damage, blindness, and deafness; it can also, of course, cause death. Humans can protect themselves from contaminated food to some extent by not consuming it, but the whale has no such choice. The chemicals we dump into our oceans may eventually cause such damage to whale populations that they will be unable to maintain their numbers, or perhaps even to survive.

If whales can somehow adapt to the poisons they are consuming, they still face the threat of disruption by an increasing human population, especially along coastal waters and on once remote islands. The humpback is especially vulnerable because of its shallow-water breeding and mating habits. In the past ten years, whale-watching has become big business in some humpback breeding and feeding grounds, and boat traffic of all kinds has increased throughout the humpback's territory (see Fig. 44). No one knows for certain what damage such activities may do.

Because the humpback is extremely vocal, especially during the breeding and mating season, there is some concern that increasing noise from ship traffic will adversely affect it. Researchers are now studying the effects of high-pitched underwater noise produced by hydrofoils traveling daily among breeding humpbacks in Hawaii. Recently Japanese scientists have suggested that increasing ship traffic has disturbed the migration routes of the Minke whale and Baird's beaked whale in the Pacific. Ship traffic and offshore human activity will unquestionably increase in both the feeding and mating/calving grounds of the humpback. It will be important to carry out careful population counts to keep track of the effects of these disturbances.

One way to prevent the harassment of whales is to create sanctuaries in which they can be allowed to mate, calve, or feed without being disturbed. A humpback sanctuary already exists off Maui, Hawaii. Others have been proposed at Campbell Island, New Zealand; in Tonga; on the southeast and east coasts of Australia; in the northeast Pacific off the United States and Canada; and on Silver Bank in the West Indies.

During the 1970's whale research reached a peak throughout the world. Although many questions have

been answered, many mysteries still surround the whale and its life in the oceans. Today it is difficult to fund cruises to go out and study whales in the wild, and studies involving whales near shore are becoming more difficult to conduct because of interference and disturbance from nonscientists. The creation of whale sanctuaries would not only protect the whale but give scientists an opportunity to study whales undisturbed in their natural environment. We think this is an idea whose time has come.

The humpback whale has survived for ten million years. It has lived through climatic changes and has been hunted and slaughtered almost to extinction. There is some evidence that its numbers are at last increasing, but as this chapter has made clear serious problems remain. Continued advances in techniques of study such as radio tagging, underwater observation, individual identification and the ability to sex animals at sea, will help in our efforts to assure the humpback's survival. The future is not ours to know, but we may hope that Bullen's visionary observations of 1904 become a reality for the humpbacks of the future:

Amiable, fondest of parents, content to play about the shores of the most beautiful islands in the world, and immune from attack of man everywhere. . . . So that the joyous Humpback is practically free to enjoy his life, to eat and love and play in the vastest playground given by God.

Fig. 44. A whale-watcher tries to make contact with a humpback whale. *Photo by J. Michael Williamson.*

BIBLIOGRAPHY

Acousta, J. *The Natural and Moral History of the Indies*. Reprinted from trans. of Edward Grimston (revised by Clements R. Markham), London: Hakluyt Society, 1604.

Adams, J. E. Marine industries of the St. Vincent Grenadines, West-Indies. Ph.D. Thesis, University of Minnesota, 1970.

———. Historical geography of whaling in Bequia Island, West Indies. *Caribbean Studies*, Vol. 2, No. 3, University of Puerto Rico, 1975.

Aldrich, H. L. *Arctic Alaska and Siberia, or Eight Months with the Arctic Whalemen*. Chicago: Rand McNally, 1899.

Allen, G. M. *The Whalebone Whales of New England*. Boston: Memoirs of the Boston Society of Natural History, 1916.

Allen, J. A. Catalogue of the mammals of Massachusetts. *Bulletin Museum of Comparative Zoology*, Vol. 1, No. 8:143–252, 1869.

Alpers, A. *Dolphins, the Myth and the Mammal*. Boston: Houghton Mifflin Co., 1961.

Andrews, R. C. Observations on habits of the finback and humpback whales of the eastern North Pacific. *Bull. Amer. Mus. of Natural History*, Vol. 26:213–226, 1909.

Baker, C. S., P. H. Forestell, and R. C. Antinoja. Interactions of the Hawaiian humpback whale, *Megaptera novaeangliae*, with the right whale, *Balaena glacialis*, and odontocete cetaceans. Abstract, Third Biennial Conference on the Biology of Marine Mammals, Seattle, 1979.

Baker, C. S., and L. M. Herman. Seasonal contrasts in the social behavior of the North Pacific humpback whale. Abstract, Fifth Biennial Conference on the Biology of Marine Mammals, The Society for Marine Mammalogy, Boston, 1983.

Baker, C. S., L. M. Herman, B. G. Bays, and G. B. Bauer. The impact of vessel traffic on the behavior of humpback whales. Abstract, Fifth Biennial Conference on the Biology of Marine Mammals, The Society for Marine Mammalogy, Boston, 1983.

Baker, C. S., L. M. Herman, W. Stifel, B. G. Bays, and A. Wallman. The migratory movement of humpback whales between Hawaii and Alaska. Abstract, Fifth Biennial Conference on the Biology of Marine Mammals, The Society for Marine Mammalogy, Boston, 1983.

Bauer, G. B. A., M. Fuller, J. R. Dunn, J. Zolger, and L. M. Herman. Biomagnetic studies of cetaceans. Abstract, Fifth Biennial Conference on the Biology of Marine Mammals, The Society for Marine Mammalogy, Boston, 1983.

Beamish, P. Evidence that a captive humpback whale (*Megaptera novaeangliae*) does not use sonar. *Deep-Sea Res.*, Vol. 25, No. 5:469–472, 1978.

———. Behavior and significance of entrapped baleen whales. In *The Behavior of Marine Animals, Vol. 3: Cetaceans*, ed. H. E. Winn and B. L. Olla, pp. 291–309, New York: Plenum Press, 1979.

BIBLIOGRAPHY

Beamish, P., and S. Carrol. Behavior of a humpback whale, before and after the attachment of a satellite tracking tag. Abstract, Fifth Biennial Conference on the Biology of Marine Mammals, The Society for Marine Mammalogy, Boston, 1983.

Beck, H. *Folklore and the Sea*. Middletown, Connecticut: Wesleyan Univ. Press, 1973.

Beddard, F. E. *A Book of Whales*. New York: G. P. Putnam's Sons, 1900.

Bennett, F. D. *Narrative of a Whaling Voyage Round the Globe from Year 1833–1836*. London: Richard Bentley, 1840.

Bonnaterre, A. Tableau encyclopédique et méthodique des trois regnes de la nature, dédié et présenté à M. Necker, Ministre d'Etat, et Directeur Général des Finances. *Cétologie*, No. 4:1–28, Paris, 1789.

Bredin, K. Foraging ecology of humpback whales off Newfoundland. Abstract, Fifth Biennial Conference on the Biology of Marine Mammals, The Society for Marine Mammalogy, Boston, 1983.

Breiwick, J. M., E. Mitchell, and R. R. Reeves. Simulated population trajectories for northwest Atlantic humpback whales, 1865–1980. Abstract, Fifth Biennial Conference on the Biology of Marine Mammals, The Society for Marine Mammalogy, Boston, 1983.

Brinkman, A. The identification and names of our fin whale species. *Norsk Hvalfangst-Tidende*, Vol. 56, No. 3:49–53, 1967.

Brown, R. On the history and geographical relations of the cetacea frequenting Davis Strait and Baffins Bay. *Proc. Zool. Soc.*, No. 35:533–556, 1868.

Brown, S. G. Whales observed in the Atlantic Ocean, notes on their distribution. *The Mar. Observer*, Oct. 1958.

Bryant, P. J., G. Nichols, T. B. Bryant, and K. Miller. Krill availability and the distribution of humpback whales in Southeastern Alaska. *Jour. Mamm.*, Vol. 62, No. 2:427–430, 1981.

Bryant, P., J. Perkins, G. Nichols, and D. Patten. Population status and ecology of humpback whales off West Greenland. Abstract, Fourth Biennial Conference on the Biology of Marine Mammals, San Francisco, 1981.

Bullen, F. T. *Denizens of the Deep*. London: Fleming H. Revell Co., 1904.

Carlson, C. A., and C. A. Mayo. Changes in the pigment and scar patterns on the ventral surface of the flukes of humpback whales observed in the waters of Stellwagen Bank, MA. Abstract, Fifth Biennial Conference on the Biology of Marine Mammals, The Society for Marine Mammalogy, Boston, 1983.

Chittleborough, R. G. Aerial observations on the humpback whale *Megaptera nodosa* (Bonnaterre) with notes on other species. *Australian Jour. Mar. Freshwater Res.*, Vol. 4, No. 2:219–226, 1953.

———. Studies on the ovaries of the humpback whale *Megaptera nodosa* (Bonnaterre), on the western Australian coast. *Australian Jour. Mar. Freshwater Res.*, Vol. 5, No. 1:35–62, 1954.

———. Aspects of reproduction in the male humpback whale *Megaptera nodosa* (Bonnaterre). *Australian Jour. Mar. Freshwater Res.*, Vol. 6, No. 1:1–29, 1955.

———. Puberty, physical maturity, and relative growth of the female humpback whale *Megaptera nodosa* (Bonnaterre) on the western Australian coast. *Australian Jour. Mar. Freshwater Res.*, Vol. 6, No. 3:315–27, 1955.

———. The breeding cycle of the breeding humpback whale

Megaptera nodosa (Bonnaterre). *Australian Jour. Mar. Freshwater Res.*, Vol. 9, No. 1:1–18, 1958.

———. Australian marking of humpback whales. *Norsk Hvalfangst-Tidende*, Vol. 48, No. 2:47–55, 1959.

———. Intermingling of two populations of humpback whales. *Norsk Hvalfangst-Tidende*, Vol. 48, No. 10:510–521, 1959.

———. Determination of age in the humpback whale *Megaptera nodosa* (Bonnaterre). *Australian Jour. Mar. Freshwater Res.*, Vol. 10, No. 2:125–143, 1959.

Colenso, W. *Ancient Tidelore and Tales of the Sea From the Two Ends of the World.* Napier, New Zealand: Haukes Bay Philisophical Inst., 1887.

Cope, E. D. *Megaptera bellicosa. Proceed. Amer. Philosophical Soc.*, Vol. 12:103–108, 1871.

Dakin, W. J. *Whaleman Adventurers: The Story of Whaling in Australian Waters and Other Southern Seas Related Thereto, From the Days of Sails to Modern Times.* Sydney: Angus & Robertson, 1934.

Darling, J. D. Mating behavior of "Hawaiian" humpback whales (*Megaptera novaeangliae*). Abstract, Fifth Biennial Conference on the Biology of Marine Mammals, The Society for Marine Mammals, Boston, 1983.

Darling, J. D., K. M. Gibson, and G. K. Silber. Observations on the abundance and behavior of humpback whales (*Megaptera novaeangliae*) off West Maui, Hawaii 1977–79. In *Communication and Behavior of Whales*, ed. R. Payne, AAAS Selected Symposium No. 76, pp. 201–222, Boulder, Colorado: Westview Press, 1983.

Darling, J. D., and C. M. Jurasz. Migratory destinations of North Pacific humpback whales (*Megaptera novaeangliae*). In *Communication and Behavior of Whales*, ed. R. Payne,

AAAS Selected Symposium No. 76, pp. 359–368, Boulder, Colorado: Westview Press, 1983.

Darling, J. D., and D. J. McSweeney. Observations on the migrations of North Pacific humpback whales (*Megaptera novaeangliae*). Abstract, Fifth Biennial Conference on the Biology of Marine Mammals, The Society for Marine Mammalogy, Boston, 1983.

Dawbin, W. H. Whale marking in South Pacific waters. *Norsk Hvalfangst-Tidende*, Vol. 45, No. 9:485–508, 1956.

———. Movements of humpback whales marked in the south west Pacific Ocean 1952–1962. *Norsk Hvalfangst-Tidende*, Vol. 53, No. 3:68–78, 1964.

———. The migration of humpback whales which pass the New Zealand coast. *Trans. R. Soc. N.Z.*, Vol. 84:147–196, 1956.

———. World stocks of humpback whales. Unpublished paper, Advisory Committee on Marine Resources Research, Norway, 13 Aug.–9 Sept. 1976.

———. The seasonal migratory cycle of humpback whales. *Whales, Dolphins and Porpoises*, ed. K. S. Norris, pp. 145–170, Berkeley: Univ. of California Press, 1966.

Delymure, S. L. Helminthofauna of marine mammals (ecology and phylogeny). Moscow: Academy of Sciences of the USSR, 1955. (Jerusalem: Israel Program for Scientific Translations, 1968.)

Dewhurst, H. W. *The Natural History of the Order Cetacea.* London, 1834.

Dohl, T. P. Return of the humpback whale (*Megaptera novaeangliae*) to central California. Abstract, Fifth Biennial Conference on the Biology of Marine Mammals, The Society for Marine Mammalogy, Boston, 1983.

Dolphin, W. F., and D. McSweeney. Aspects of the foraging strategies of humpback whales determined by hydro-acoustic scans. Abstract, Fourth Biennial Conference on the Biology of Marine Mammals, San Francisco, 1981.

Dudley, P. An essay upon the natural history of whales. *Philos. Trans.*, Vol. 33, No. 387:256–269, 1725.

Dudley, Sir R. *The Voyage of Sir Robert Dudley to the West Indies, 1594–1595*. London: Hakluyt Society, 1899.

Dudok van Heel, W. H. Sound and cetacea. *Netherlands Jour. of Sea Res.*, Vol. 60, No. 1:407–507, 1977.

DuTetre, J. B. *Histoire Général des Antilles*. Tom. 2, Traité 4, "Des Poissons," 1667.

Earle, S. A. Quantitative sampling of krill (*Euphausia pacifica*) relative to feeding strategies of humpback whales (*Megaptera novaeangliae*) in Glacier Bay, Alaska. Abstract, Third Biennial Conference on the Biology of Marine Mammals, Seattle, 1979.

Edel, R. K., and H. E. Winn. Observations on the underwater locomotion and flipper movement of the humpback whales. *Mar. Biol.*, Vol. 48:279–287, 1978.

Edwards, E. J., and J. Edwards Rattray. *Whale Off! The Story of American Shore Whaling*. New York: Coward-McCann Inc., 1932.

Ekman, Sven. *Zoogeography of the Sea*. London: Sidgwick and Jackson, 1953.

Fabricius, O. Zoologiske bidrag. *K. Danske Videns.-Selsk. Skrivter*, No. 6:63–83, 1818.

Fanning, J. The blow of whales—a functional explanation. Abstract, Third Biennial Conference on the Biology of Marine Mammals, Seattle, 1979.

Fenger, F. A. Longshore whaling in the Grenadines. *Outing Magazine*, No. 61:664–679, 1913.

Fish, M. P. Marine mammals of the Pacific with particular ref-erence to the production of underwater sound. In *Tech. Rept. No. 8*, ed. C. J. Fish, Office of Naval Research, 1949.

Fleischer, G. Hearing in extinct cetaceans as determined by cochlear structure. *Jour. Paleontology*, Vol. 50, No. 1:133–152, 1976.

Flower, W. H. *Recent Memoirs on the Cetacea by Professors Eschricht, Reinhardt, and Lilljeborg*. London: Robert Hardwicke, 1866.

Food and Agriculture Organization. *Mammals in the Sea, Vol. I*. FAO Fish. Series No. 5, Rome, 1978.

Forestell, P. H., and L. M. Herman. Behavior of "escort" accompanying mother-calf pairs of humpback whales. Abstract, Third Biennial Conference on the Biology of Marine Mammals, Seattle, 1979.

Frumhoff, P. Aberrant songs of humpback whales (*Megaptera novaeangliae*): clues to the structure of humpback songs. In *Communication and Behavior of Whales*, ed. R. Payne, AAAS Selected Symposium No. 76, pp. 81–127, Boulder, Colorado: Westview Press, 1983.

Fujino, K. On the blood groups of the sei, fin, blue and humpback whales. *Proc. Japanese Academy*, Vol. 29, No. 4:182–191, 1962.

Gambell, R. Some effects of exploration on reproduction in whales. *Jour. Reprod. Fert.*, Suppl., Vol. 19:533–553, 1973.

Gates, Sir Thomas, Sir George Sommerson, and Captain Newport. *A Discovery of the Bermudas, Otherwise Called the Isle of Devils*. Set forth by Fil. Jourdan in Richard Hakluyt, *Voyages and Discoveries of the English*, London, 1812.

Gersh, I. Note on pineal gland of humpback whale. *Jour. Mamm.*, Vol. 19:477–480, 1938.

Glockner, D. A. Determining the sex of humpback whales

(*Megaptera novaeangliae*) in their natural environment. In *Communication and Behavior of Whales*, ed. R. Payne, AAAS Selected Symposium No. 76, pp. 447–464, Boulder, Colorado: Westview Press, 1983.

Glockner, D. A., and S. C. Venus. Humpback whale (*Megaptera novaeangliae*) cows with calves identified off West Maui, Hawaii, 1977–78. Abstract, Third Biennial Conference on the Biology of Marine Mammals, Seattle, 1979.

———. Identification, growth rate, and behavior of humpback whale (*Megaptera novaeangliae*) cows and calves in the waters off Maui, Hawaii, 1977–79. In *Communication and Behavior of Whales*, ed. R. Payne, AAAS Selected Symposium No. 76, pp. 223–258, Boulder, Colorado: Westview Press, 1983.

Glockner-Ferrari, D., and M. Ferrari. Reproduction, aggression, and sexual activities in the humpback whale. Abstract, Fifth Biennial Conference on the Biology of Marine Mammals, The Society for Marine Mammalogy, Boston, 1983.

———. Correlation of the sex and behavior of individual humpback whales *Megaptera novaeangliae* to their role in the breeding population. Abstract, Fourth Biennial Conference on the Biology of Marine Mammals, San Francisco, 1981.

Goodale, T. B. With the whalers at Durban, and a few notes on the anatomy of the humpback whale. (*Megaptera boops*). *The Zoologist*, Vol. 17, No. 864:201–211, 1913.

Goode, G. B. *The Fisheries and Fishing Industries of the U.S.* Washington, D.C.: U.S. Govt. Printing Office, 1887.

Goodyear, J. D. "Remora" tag effects, the first radio-tracking of an Atlantic humpback. Abstract, Fourth Biennial Conference on the Biology of Marine Mammals, San Francisco, 1981.

———. Night behavior of humpback whales in the Gulf of Maine as determined by radio tracking. Abstract, Fifth Biennial Conference on the Biology of Marine Mammals, The Society for Marine Mammalogy, Boston, 1983.

Gray, J. E. On the Bermuda humpbacked whale of Dudley (*Balaena nodosa*, Bonnaterre, *Megaptera americana*, Gray, and *Megaptera bellicosa*, Cope). *Ann. and Mag. Nat. Hist.*, Vol. 13, No. 74:186, 1874.

Gregory, Lady. *A Book of Saints and Wonders Put Down Here by Lady Gregory According to the Old Writings and the Memory of the People of Ireland*. London: John Murray, 1908.

Guinee, L. N., K. Chu, and E. M. Dorsey. Changes over time in the songs of known individual humpback whales (*Megaptera novaeangliae*). In *Communication and Behavior of Whales*, ed. R. Payne, AAAS Selected Symposium No. 76, pp. 59–80, Boulder, Colorado: Westview Press, 1983.

Hafner, G. W., C. L. Hamilton, W. W. Steiner, T. J. Thompson, and H. E. Winn. Evidence for signature information in the song of the humpback whale. *Jour. Acoust. Soc. Am.*, Vol. 66, No. 1:1–6, 1979.

Hain, J. H. W., G. R. Carter, S. D. Kraus, C. A. Mayo, and H. E. Winn. Feeding behavior of the humpback whale, *Megaptera novaeangliae*, in the western North Atlantic. *Fish. Bull.*, Vol. 80, No. 2:259–268, 1982.

Hakluyt, R. *The Principal Navigations, Voyages, Traffiques and Discoveries of the English Nation*. London, 1599.

Haldane, R. C. Notes on whaling in Shetland. *The Annals of Scottish Natural History*, pp. 54–72, 1905.

Hamilton, R. *Mammalia, Vol. VI, On the Ordinary Cetacea or Whales*. Edinburgh, 1839.

Herman, L. M. Humpback whales in the breeding waters:

population and pod characteristics. *Sci. Rept. Whales Res. Inst.*, No. 29:59–85, 1977.

———. *Humpback Whales in Hawaiian Waters: A Study in Historical Ecology*. Seattle: Pacific Search Press, 1978.

———, ed. *Cetacean Behavior: Mechanisms and Functions*. New York: John Wiley & Sons, 1980.

Herman, L. M., R. C. Antinoja, C. S. Baker, and R. S. Wells. Temporal and spatial distribution of humpback whales in Hawaii. Abstract, Third Biennial Conference on the Biology of Marine Mammals, Seattle, 1979.

Ingebrigtsen, A. Whales caught in the north Atlantic and other seas. *Rapp. P.-v. Réun. Cons. Perm. Int. Explor. Mer*, Vol. 56:1–26, 1929.

Johnson, J. H. Glacier Bay humpback whale research: what have we learned? Abstract, Fifth Biennial Conference on the Biology of Marine Mammals, The Society for Marine Mammalogy, Boston, 1983.

Jones, E. C. A squaloid shark, the probable cause of crater wounds on fishes and cetaceans. *Fish. Bull.*, Vol. 69, No. 4:791–798, 1971.

Jones, J. M. Mammals of Bermuda. *Bull. U.S. National Museum*, No. 25:143–161, 1884.

Jurasz, C. M., and V. P. Jurasz. Feeding modes of the humpback whale, *Megaptera novaeangliae*, in southeast Alaska. *Sci. Rep. Whales Res. Inst.*, No. 31:67–81, 1979.

Katona, S. K., K. C. Balcomb III, J. A. Beard, H. Whitehead, and D. Matilla. The Atlantic humpback whale catalogue. Abstract, Fifth Biennial Conference on the Biology of Marine Mammals, The Society for Marine Mammalogy, Boston, 1983.

Katona, S. K., B. Baxter, O. Brazier, J. Perkins, and H. Whitehead. Identification of humpback whales by fluke photographs. In *Behavior of Marine Animals, Vol. 3: Cetaceans*, ed. H. E. Winn and B. Olla, pp. 33–44, New York: Plenum Press, 1979.

Katona, S. K., P. Harcourt, J. S. Perkins, and S. D. Kraus, eds. *Humpback Whales: A Catalog of Individuals Identified in the Western North Atlantic Ocean by Means of Fluke Photographs*. Bar Harbor, Maine: College of the Atlantic, 1980.

Kaufman, G. Ecology of humpback whales in American Samoa. Abstract, Fifth Biennial Conference on the Biology of Marine Mammals, The Society for Marine Mammalogy, Boston, 1983.

Kaufman, G., and K. Wood. Demographics of humpback whales off southwest Maui, Hawaii. Abstract, Fifth Biennial Conference on the Biology of Marine Mammals, The Society for Marine Mammalogy, Boston, 1983.

Kennett, J. P. Development of planktonic biogeography in the southern ocean during Cenozoic. *Mar. Micropaleontology*, Vol. 3:301–345, 1978.

Kennett, J. P., R. E. Burns, J. E. Andrews, M. Churkin, T. A. Davies, P. Dumitrica, A. R. Edwards, J. S. Glaehouse, G. H. Packham, and G. J. van de Tingen. Australian and Antarctic continental drift, palaeo-circulation changes and Oligocene deep-sea erosion. *Nature Physical Sci.*, Vol. 239, No. 91:51–55, 1972.

Kenney, R. D. Distributional biology of the cetacean fauna of the northeast United States continental shelf. Ph.D. Thesis, University of Rhode Island, 1984.

Kenney, R. D., D. R. Goodale, G. P. Scott, and H. E. Winn. Spatial and temporal distribution of humpback whales in the CETAP study area. In *A Characterization of Marine Mammals and Turtles in the Mid- and North Atlantic Areas of the U.S. Outer Continental Shelf: Annual Report for*

1979, ed. H. E. Winn, Washington, D.C.: Bureau of Land Management, 1980 (Ref. No. AA551-CT8-48).

Kenney, R. D. and H. E. Winn. An hypothesis accounting for differences in fluking behavior in whales. Abstract, Fifth Biennial Conference on the Biology of Marine Mammals, The Society for Marine Mammalogy, Boston, 1983.

Klinowska, M. Is the cetacean map geomagnetic? Evidence from strandings. Abstract, *Aquatic Mammals*, Vol. 10:17, 1983.

Kraus, S. Whale-fishermen interactions in Newfoundland, a review of the problem and potential solutions. *Echo* (College of the Atlantic's Research Log, Bar Harbor, Maine), Vol. 4, No. 2, 1977.

Lattimore, R. *The Odyssey of Homer*. New York: Harper & Row, 1967.

Leatherwood, S., and W. E. Evans. Some recent uses and potentials of radiotelemetry in field studies of cetaceans. In *Behavior of Marine Animals, Vol. 3: Cetaceans*, ed. H. E. Winn and B. Olla, pp. 1–31, New York: Plenum Press, 1979.

Levenson, C., and W. T. Leapley. Humpback whale distribution in the eastern Caribbean determined acoustically from an oceanographic aircraft. NAVOCEANO Tech. Note 3700-46-76, Washington, D.C.: U.S. Naval Oceanographic Office, 1976.

Lien, J. A study of whale entrapment in fishing gear: causes and prevention. Unpublished progress report, Memorial University, St. John's, Newfoundland, 1979.

Lien, J., and H. Whitehead. Changes in humpback (*Megaptera novaeangliae*) abundance off northeast Newfoundland related to the status of capelin (*Mallotus villosus*) stocks. Abstract, Fifth Biennial Conference on the Biology of Marine

Mammals, The Society of Marine Mammalogy, Boston, 1983.

Lipps, J. H., and E. Mitchell. Trophic model for the adaptive radiations and extinctions of pelagic marine mammals. *Paleobiology*, Vol. 2, No. 2:147–155, 1976.

Macy, O. *History of Nantucket*. Boston: Hilliard, Gray, and Co., 1835.

Martin, A. R., S. K. Katona, D. Matilla, D. Hembree, and T. D. Waters. Migration of humpback whales between the Caribbean and Iceland. *Jour. Mamm.*, Vol. 65, No. 2: 333–336, 1984.

Mate, B. R. Movements and dive characteristics of a satellite monitored humpback whale. Abstract, Fifth Biennial Conference on the Biology of Marine Mammals, The Society for Marine Mammalogy, Boston, 1983.

Matilla, D. K. Humpback whales off Puerto Rico: population composition and habitat use. Abstract, Fifth Biennial Conference on the Biology of Marine Mammals, The Society for Marine Mammalogy, Boston, 1983.

Matthews, L. H. The humpback whale *Megaptera nodosa*. *Discovery Reports*, Vol. 17:7–92, 1938.

Matthews, L. H. *The Whale*. New York: Crescent Books, 1973.

Mayo, C. A. Patterns of distribution and occurrence of humpback whales in the southern Gulf of Maine. Abstract, Fifth Biennial Conference on the Biology of Marine Mammals, The Society for Marine Mammalogy, Boston, 1983.

Mayo, C. A., and P. J. Clapham. Observations of humpback whale mother/calf pairs on Stellwagen Bank, MA: 1979–83. Abstract, Fifth Biennial Conference on the Biology of Marine Mammals, The Society for Marine Mammalogy, Boston, 1983.

BIBLIOGRAPHY

McSweeney, D., W. Dolphin, and R. Payne. Humpback whale (*Megaptera novaeangliae*) songs recorded on summer feeding grounds. Abstract, Fifth Biennial Conference on the Biology of Marine Mammals, The Society for Marine Mammalogy, Boston, 1983.

Melville, H. *Moby-Dick, or the Whale*. Berkeley: Univ. of California Press, 1981.

Meyer, K., ed. and transl. *The Voyage of Bran*. 2 vols., London: D. Nutt, 1895.

Millais, J. G. *The Mammals of Great Britain and Ireland, Vol. 3: Cetacea*. London, 1906.

Mitchell, E. D. The status of the world's whales. *Nature Canada*, Vol. 2, No. 4:9–27, 1973.

Morrison, S. E. *The European Discovery of America. The Southern Voyages* A.D. *500–1600*. New York: Oxford Univ. Press, 1971.

Mörzer Bruyns, W. F. *Field Guide of Whales and Dolphins*. Uitgeverij Tor N.V. Uitgeverij V.H.C.A. Mees Zieseniskade 14, Amsterdam, Netherlands, 1971.

Nemoto, T. Feeding pattern of baleen whales in the ocean. In *Marine Food Chains*, ed. J. H. Steele, pp. 241–255, Edinburgh: Oliver & Boyd, 1970.

Nicol, C. J. A. *The Biology of Marine Animals*. London: Sir Isaac Pitman & Sons Ltd., 1960.

Nishiwaki, M. Ryukyuan whaling in 1961. *Sci. Rep. Whales Res. Inst.*, No. 16:19–28, 1962.

Nordhoff, C. *Whaling and Fishing*. Cincinnati: Moore, Wilstach, Keys & Co., 1856.

Norris, K. S. Some observations on migration and orientation of marine mammals. In *Animal Orientation and Navigation*, ed. R. M. Storm, pp. 101–126, Corvallis: Oregon State Univ. Press, 1967.

Ommanney, F. D. *Lost Leviathan*. London: Hutchinson & Co. Ltd., 1971.

Oppian. *Halieutica*, transl. A. W. Mair. Loeb Classical Library, Cambridge: Harvard Univ. Press, 1928.

Oviedo, G. de. *Historia General y Natural de las Indias*. 4 vols., Madrid: Royal Academy of History, 1851–1855.

Payne, K. Progressive changes in songs of humpback whales. Abstract, Third Biennial Conference on the Biology of Marine Mammals, Seattle, 1979.

Payne, K., P. Tyack, and R. Payne. Progressive changes in the songs of humpback whales: a detailed analysis of two seasons in Hawaii. In *Communication and Behavior of Whales*, ed. R. Payne, AAAS Selected Symposium No. 76, pp. 9–57, Boulder, Colorado: Westview Press, 1983.

Payne, P. M., K. D. Powers, J. R. Nicolas, and L. O'Brien. Apparent shift in the distribution of the humpback whale (*Megaptera novaeangliae*) in the Gulf of Maine in response to increased densities of the sand eel (*Ammodytes americanus*). Abstract, Fifth Biennial Conference on the Biology of Marine Mammals, The Society for Marine Mammalogy, Boston, 1983.

Payne, R. Behavior and vocalizations of humpback whales (*Megaptera* sp.). In *Report on a Workshop on Problems Related to Humpback Whales (Megaptera novaeangliae) in Hawaii*, ed. K. S. Norris and R. R. Reeves, Report No. MMC-77/03, Washington, D.C.: U.S. Marine Mammal Commn., 1978.

———. Humpback whale songs as an indicator of "stock." Abstract, Third Biennial Conference on the Biology of Marine Mammals, Seattle, 1979.

———, ed. *Communication and Behavior of Whales*. AAAS Selected Symposium No. 76, Boulder, Colorado: Westview

Press, 1983. (There are eight important papers on the humpback whale in this book.)

Payne, R., and L. N. Guinee. Humpback whale (*Megaptera novaeangliae*) songs as an indicator of "stocks." In *Communication and Behavioar of Whales*, ed. R. Payne, AAAS Selected Symposium No. 76, pp. 333–358, Boulder, Colorado: Westview Press, 1983.

Payne, R., and S. McVay. Songs of humpback whales. *Science*, Vol. 173, No. 3997:587–597, 1971.

Payne, R., and D. Webb. Orientation by means of long range acoustic signaling in baleen whales. *Annals New York Academy Sci.*, Vol. 188:110–141, 1971.

Perkins, J. S., K. C Balcomb III, and G. Nichols, Jr. West Greenland humpbacks, update to 1983. Abstract, Fifth Biennial Conference on the Biology of Marine Mammals, The Society for Marine Mammalogy, Boston, 1983.

Perkins, J. S., and P. C. Beamish. Net entanglements of baleen whales in the inshore fishery of Newfoundland. *Jour. Fish. Res. Board Canada*, Vol. 36, No. 5:521–528, 1979.

Rathjen, W. F., and J. R. Sullivan. West Indies whaling. Caribbean Fisheries Development Project, F.A.O., Barbados.

Rayner, G. W. Whale marking. *Discovery Reports*, Vol. 19: 245–284, 1940.

Reichley, N. E. Sound production by the humpback whale (*Megaptera novaeangliae*) in Cape Cod, MA waters. Abstract, Fifth Biennial Conference on the Biology of Marine Mammals, The Society for Marine Mammalogy, Boston, 1983.

Rice, D. W. The humpback whale in the North Pacific: distribution, exploitation and numbers. In *Report on Workshop on Problems Related to Humpback Whales in Hawaii*, ed. K. S. Norris and R. R. Reeves, Rept. No. MMC-77/03,

Washington, D.C.: U.S. Marine Mammal Commn., 1978.

Ridgway, S. H. *Mammals of the Sea: Biology and Medicine*. Springfield, Illinois: Charles C Thomas, 1972.

Risting, S. Av hvalfangstens historie. Publikation Nr. 2 fra Kommandor Chr. Christensens, Hvalfangstmuseum, 1 Sandefojord, Kristiania, 1922.

Rochefort, C. de. *Histoire Naturelle des Illes Antilles*. Rotterdam: Arnould Leers, 1658.

Sanderson, I. T. *Follow the Whale*. Boston: Little, Brown & Co., 1956.

Scammon, C. M. *The Marine Mammals of the North-Western Coast of North America and the American Whale Fishery*. San Francisco: J. H. Carmany & Co., 1874. (Reprint, Riverside, Calif.: Manessier Pub. Co., 1969.)

Schevill, W. E. Daily patrol of a *Megaptera*. *Jour. Mamm.*, Vol. 41, No. 2:279–281, 1960.

———. Underwater sounds of cetaceans. In *Marine Bioacoustics*, ed. W. N. Tavolga, pp. 307–316, New York: Pergamon Press, 1964.

———. *The Whale Problem: A Status Report*. Cambridge: Harvard Univ. Press, 1974.

Schreiber, O. W. Some sounds from marine life in the Hawaiian area. *Jour. Acoust. Soc. Am.*, Vol. 24, No. 1:116, 1952.

Scott, G. P. Aerial and shipboard estimates of humpback whale abundance on Silver and Navidad Banks, West Indies. In *Proceedings of a Workshop on Humpback Whales in the Western North Atlantic*, ed. J. H. Prescott, S. D. Kraus, and J. R. Gilbert, abstract, Woods Hole, Mass.: Natl. Marine Fisheries Service, 1980.

Scott, G. P., and H. E. Winn. Comparative evaluation of aerial and shipboard sampling techniques for estimating the abundance of humpback whales (*Megaptera novaeangliae*). Report No. MMC-79/24, NTIS Publ. PB81-109852, Washington, D.C.: U.S. Marine Mammal Commn., 1980.

———. Aerial assessment of humpback whale (*Megaptera novaeangliae*) stocks using vertical photographs. Proc. PECORA IV Symposium, National Wildlife Federation Scientific and Technical Series No. 3:235–243, 1978.

Sears, R. Humpback whale (*Megaptera novaeangliae*) distribution and dispersal in the Gulf of St. Lawrence. Abstract, Fifth Biennial Conference on the Biology of Marine Mammals, The Society for Marine Mammalogy, Boston, 1983.

Severin, T. *The Brendan Voyage*. New York: McGraw-Hill, 1978.

Shindo, N. *History of Whales in the Inland Sea*. Junnosuke Oomura, Toshie Nuguishi, 1975.

Silber, G. Social phonations and the associated surface behavior of Hawaiian humpback whales (*Megaptera novaeangliae*). Abstract, Fifth Biennial Conference on the Biology of Marine Mammals, The Society for Marine Mammalogy, Boston, 1983.

Silber, G., and D. McSweeney. A comparison of social phonations of the humpback whale (*Megaptera novaeangliae*) in Hawaii and southeast Alaska. Abstract, Fifth Biennial Conference on the Biology of Marine Mammals, The Society for Marine Mammalogy, Boston, 1983.

Slijper, E. J. *Whales*. London: Hutchinson, 1962.

Stackpole, E. A. *Whales & Destiny*. Amherst: Univ. of Massachusetts Press, 1972.

Stone, G., and S. Katona. Humpback whales of Bermuda. Abstract, Fifth Biennial Conference on the Biology of Marine Mammals, The Society for Marine Mammalogy, Boston, 1983.

Stump, C. W., J. P. Robins, and M. L. Garde. The development of the embryo and membranes of the humpback whale, *Megaptera nodosa* Bonnaterre. *Australian Jour. Mar. Freshwater Res.*, Vol. 11, No. 3:365–386, 1960.

Symons, M. A., and R. D. Weston. Studies on the humpback whale (*Megaptera nodosa*) in the Bellingshausen Sea. *Norsk Hvalfangst-Tidende*, Vol. 47, No. 2:53–81, 1958.

Taruski, A. G., C. E. Olney, and H. E. Winn. Chlorinated hydrocarbons in cetaceans. *Jour. Fish. Res. Board Canada*, Vol. 32, No. 11:2205–2209, 1975.

Thompson, T. J. Temporal characteristics of humpback whale (*Megaptera novaeangliae*) songs. Ph.D. Thesis, University of Rhode Island, 1981.

Thompson, T. J., H. E. Winn, and P. J. Perkins. Mysticete sounds. In *Behavior of Marine Animals, Vol. 3: Cetaceans*, ed. H. E. Winn and B. L. Olla, pp. 403–431, New York: Plenum Press, 1979.

Tillman, M. F. Assessment of North Pacific stocks of whales. *Mar. Fish. Rev.*, Vol. 37, No. 10:1–4, 1975.

Tomilin, A. G. *On the Behavior and Sonic Signalling of Whales*, transl. A. DeVreeze and D. E. Sergeant. Fish. Res. Board Canada, 1955.

———. *Mammals of the USSR and Adjacent Countries, Vol. IX: Cetacea*, 1957. (Translation no. 1124, Jerusalem: Israel Program for Sci. Transl., 1967.)

Townsend, C. H. The distribution of certain whales as shown by logbook records of American whaleships. *Zoologica*, Vol. 19, No. 1:1–50, 1935.

True, F. W. On the nomenclature of the whalebone whales of the tenth edition of Linnaeus's *Systema Naturae*. *Proc. U.S. National Museum*, Vol. 21, No. 1163:617–635, 1898.

————. The whalebone whales of the western North Atlantic. *Smithsonian Contrib. to Knowledge*, Vol. 33:1-332, 1904.

Tyack, P. Interactions between singing Hawaiian humpback whales and conspecifics nearby. *Behav. Ecol. Sociobiol.*, Vol. 8:105–116, 1981.

Verrill, A. E. *The Bermuda Islands. An Account of Their Scenery, Climate, Productions, Physiography, Natural History and Geology, With Sketches of Their Discovery and Early History, and the Changes in Their Flora and Fauna Due to Man.* New Haven, Conn: Conn. Acad. Sci., Trans.; private reprint, 1902.

Villiers, A. *The Quest of the Schooner Argus.* New York: Charles Scribner's Sons, 1951.

Watkins, W. A., and W. E. Schevill. Aerial observation of feeding behavior in four baleen whales: *Eubalaena glacialis, Balaenoptera borealis, Megaptera novaeangliae,* and *Balaenoptera physalus. Jour. Mamm.*, Vol. 60, No. 1:155–163, 1977.

Weinrich, M. T. Association patterns in a population of humpback whales (*Megaptera novaeangliae*). Abstract, Fifth Biennial Conference on the Biology of Marine Mammals, The Society for Marine Mammalogy, Boston, 1983.

Wheeler, J. F. G. On a humpback whale taken at Bermuda. *Proc. Zool. Soc. London*, Vol. 3, Ser. B: 37–38, 1941.

Whitehead, H., P. Harcourt, K. Ingham, and H. Clark. The migration of humpback whales past the Bay de Verde Peninsula, Newfoundland, during June and July, 1978. *Canadian Jour. Zool.*, Vol. 58, No. 5:687–692, 1980.

Williamson, G. R. Winter sighting of a humpback suckling its calf on the Grand Bank of Newfoundland. *Norsk Hvalfangst-Tidende*, Vol. 50, No. 8:335–341, 1961.

Willoughby, F. *Icthyographia.* London: Royal Society, 1685.

Winn, H. E., ed. *A Characterization of Marine Mammals and Turtles in the Mid- and North Atlantic Areas of the U.S. Outer Continental Shelf: Annual Report for 1979.* Washington, D.C.: Bureau of Land Management, 1980 (Ref. No. AA551-CTG8-48).

————. *A Characterization of Marine Mammals and Turtles in the Mid- and North Atlantic Areas of the U.S. Outer Continental Shelf: Annual Report for 1980.* Washington, D.C.: Bureau of Land Management, 1982 (Ref. No. AA551-CT8-48).

————. *A Characterization of Marine Mammals and Turtles in the Mid- and North Atlantic Areas of the U.S. Outer Continental Shelf: Final Report.* Washington, D.C.: Bureau of Land Management, 1982 (Ref. No. AA551-CT8-48).

Winn, H. E., P. Beamish, and P. J. Perkins. Sounds of two entrapped humpback whales (*Megaptera novaeangliae*) in Newfoundland. *Mar. Biol.*, Vol. 55:151–155, 1979.

Winn, H. E., W. L. Bischoff, and A. G. Taruski. Cytological sexing of cetacea. *Mar. Biol.*, Vol. 23:343–346, 1973.

Winn, H. E., R. K. Edel, and A. G. Taruski. Population estimate of the humpback whale in the West Indies by visual and acoustic techniques. *Jour. Fish. Res. Board Canada*, Vol. 32, No. 4:499–506, 1975.

Winn, H. E., P. J. Perkins, and T. Poulter. Sounds of the humpback whale. In *Proc. 7th Annual Conf. Biol. Sonar*, Vol. 7:39–52, Stanford Research Institute, Menlo Park, Calif., 1970.

Winn, H. E., and D. Rice. Humpback whales in the Sea of Cortez. *Currents*, Vol. 2, No. 45:5–7, 1978.

Winn, H. E., and G. P. Scott. Evidence for three substocks of humpback whales (*Megaptera novaeangliae*) in the western North Atlantic. Intern. Council Explor. Sea, document C.M.1977/N:13, 1977, and abstract, Second Conference on the Biology of Marine Mammals, San Diego, 1977.

————. The humpback whale: present knowledge and future trends in research with special reference to the western North Atlantic. In *Mammals in the Sea, Vol. III*, FAO Fish. Series No. 5, pp. 171–180, Rome, 1981.

Winn, H. E., and T. J. Thompson. Comparison of humpback whale sounds across the northern hemisphere. Abstract, Third Biennial Conference on the Biology of Marine Mammals, Seattle, 1979.

Winn, H. E., T. J. Thompson, W. C. Cummings, J. H. W. Hain, J. Hudnall, H. E. Hays, and W. W. Steiner. Song of the humpback whale—population comparisons. *Behav. Ecol. Sociobiol.*, Vol. 8:41–46, 1981.

Winn, H. E., and L. K. Winn. Report on the status of the humpback whale on Silver, Navidad and Mouchoir Banks in the West Indies and recommendations to ensure their conservation. Washington, D.C.: U.S. Marine Mammal Commn., 1977.

————. The song of the humpback whale (*Megaptera novaeangliae*) in the West Indies. *Mar. Biol.*, Vol. 47:97–114, 1978.

Winship, G. P. *Sailors' Narratives of Voyages Along the New England Coast 1524–1624*. Boston: Houghton Mifflin, 1905.

Wolman, A., and C. Jurasz. Humpback whales in Hawaii: vessel census 1976. *Mar. Fish. Rev.*, Vol. 39, No. 71:1–7, 1977.

Yablokov, A. V., and G. A. Klevazal. *Whiskers of Whales and Seals and Their Distribution, Structure, and Significance.* Fish. Res. Board Can. Translation Series, No. 1335, 1969.

INDEX

INDEX